Legacies of the
Manhattan Project

Hanford Histories

Volume 2

Michael Mays, Series Editor

Titles in the series:
Nowhere to Remember: Hanford, White Bluffs, and Richland to 1943 (2018),
edited by Robert Bauman and Robert Franklin

Legacies of the Manhattan Project

Reflections on 75 Years of a Nuclear World

Edited by
Michael Mays

WSU
PRESS

Washington State University Press
Pullman, Washington

Washington State University Press
PO Box 645910
Pullman, Washington 99164-5910
Phone: 800-354-7360
Fax: 509-335-8568
Email: wsupress@wsu.edu
Website: wsupress.wsu.edu

© 2020 by the Board of Regents of Washington State University
All rights reserved
First printing 2020

Printed and bound in the United States of America on pH neutral, acid-free paper. Reproduction or transmission of material contained in this publication in excess of that permitted by copyright law is prohibited without permission in writing from the publisher.

Library of Congress Cataloging-in-Publication Data

Names: Mays, Michael, editor.
Title: Legacies of the Manhattan project : reflections on 75 years of a
  nuclear world / edited by Michael Mays.
Description: Pullman, Washington : Washington State University Press, 2020.
  | Series: Hanford histories; volume 2 | Includes bibliographical
  references and index.
Identifiers: LCCN 2019049575 | ISBN 9780874223750 (trade paperback)
Subjects: LCSH: Manhattan Project (U.S.)--History. | Atomic bomb--United
  States--History.
Classification: LCC QC773.3.U5 L44 2020 | DDC 355.8/25119097309044--dc23
LC record available at https://lccn.loc.gov/2019049575

Chapter 10 credits: Quote from *Nuclear Fear : A History of Images* by Spencer R. Weart, Cambridge, Mass.: Harvard University Press, Copyright © 1988 by the President and Fellows of Harvard College.

Quote from *The End of Victory Culture* by Tom Englehardt © 1995.
Reprinted by permission of Basic Books, an imprint of Hachette Book Group, Inc.

For Colleen French

For her vision, tenacity, and unwavering commitment to the preservation
of Hanford's extraordinary history

Contents

Illustrations

Acknowledgments

Volume Two of the Hanford Histories series, and indeed the series itself, would not have been possible without the support of countless individuals who have had a hand in its coming to fruition. Foremost among those who need to be acknowledged are our colleagues at Washington State University Press: retired Editor-in-Chief Robert Clark, current Editor-in-Chief Linda Bathgate, Director Edward Sala, Beth Deweese, Caryn Lawton, Kerry Darnall, and its editorial board. Their collective professionalism and their enthusiastic commitment to the mostly thankless job they do is an inspiration to all of us who work with them, gratefully.

The Tri-Cities community in Southeastern Washington is an extraordinary one. Its citizens are passionate about our region's unique history, and they amaze and challenge those of us with the Hanford History Project (HHP) daily. The depth of their knowledge is both humbling and stimulating. Our gratitude to Colleen French for her singular leadership, resourcefulness, efforts, and support on behalf of the historical preservation of our community is registered in the book's dedication. Likewise, Gary Petersen, Don Sorenson, Tom Marceau, Sharon Holden, Ann Roseberry, Diahann Howard, C. Mark Smith, Maynard Plahuta, and John Fox are each forces of nature in their own right. The Tri-Cities historical landscape, and in particular that of the Hanford Site, would be a barren one without their seemingly inexhaustible labors. HHP has been the fortunate beneficiary of close collaborations with a broad range of stellar community organizations and partners. Special thanks are due to Vanessa Moore, Leonard Moore, Vanis Daniels, and Tanya Bowers with the African American Community Cultural and Educational Society; Wayne Glines and Ron Kathren at the Herbert M. Parker Foundation; the executive board and members of the B Reactor Museum Association (BRMA); Tom Hungate and Linda Pasch with Northwest Public Broadcasting; Darci Teel, Mary Petrich-Guy, and Aaron Fergusson at Mission Support Alliance; Becky Burghart and Kris Kirby with the Manhattan Project National Historical Park Site; and Tracy Atkins from the Department of Energy's Office of Legacy Management. All were instrumental in the success of the Legacies of the Manhattan Project conference and in helping sustain HHP from its beginning. Thanks, too, to Del Ballard from BRMA and to Congressman Doc Hastings for generously having agreed to participate in the conference's opening night panel.

There would have been no conference without the enthusiastic participation of some seventy-plus presenters who traveled to our remote neck of the woods from across the United States and from six countries. Many of the essays in this volume were first offered as working papers at the Legacies conference. The remarkably rich and wide-ranging exchanges stemming from those presentations sparked the idea of this book. To those whose contributions appear here, thank you for your patience, your suggestions and insights and, most importantly, for the meticulous research you have graciously allowed us to publish here. Dr. Bryan C. Taylor and a second anonymous reviewer provided thoughtful, thorough, and most helpful comments and suggestions on the entire manuscript in an earlier form. The collection is exponentially better than it would have been without their valuable feedback.

Washington State University Tri-Cities has been an accommodating home to the Hanford History Project and for that we wish to express our gratitude to Chancellor Sandra Haynes and Vice Chancellor for Academic Affairs Kate McAteer. Dr. Robert Bauman was a founding member of HHP and inaugurated its Oral History Program. Assistant Director Robert Franklin and Program Coordinator Jillian Gardner-Andrews *are* HHP. To them, and to our student interns and our volunteers whose efforts happen mainly behind the scene, heartfelt thanks for your labors of love and your continuing support.

Finally, thanks to those current and former colleagues and collaborators at Washington State University and the University of Southern Mississippi from whom I've learned so much over the years in the classroom, in conversation, and in our shared research; to Bruce and Kerry Valentine for their exceptional friendship; to my family, Tom, Claudia, Jeff, Karen, Derek, and Daniel; and, especially, to Robin, Mila Jane, and Lennox. You are my little universe.

Introduction

Michael Mays

The impetus for this collection was the Hanford History Project's "Legacies of the Manhattan Project at 75 Years" conference, where many of the essays included here were first presented. Held in March 2017 in Richland, Washington, the conference venue is a stone's throw from the southern-most edge of the Hanford Nuclear Site. That location is where the plutonium was produced, beginning in September 1944, that fueled the "Fat Man" nuclear bomb dropped on Nagasaki, Japan, on August 9, 1945. It is a truism of scholarship that the greater the historical event—which is to say the more significant, far-reaching, and impactful it is—the greater the temporal distance needed to undertake the type of disinterested understanding, evaluation, and judgment that historians strive (however idealistically) to achieve. As one would expect, then, the scholarship that has grown up around the subject of the Manhattan Project has been accelerating, especially since the landmark publication of Richard Rhodes' seminal study, *The Making of the Atomic Bomb* (1986), and has become vast and impressive. With the 75th anniversary of its beginning at hand, the time seemed appropriate to revisit and reevaluate the project's varied and complex afterlives. And indeed, "Legacies" would certainly have been a much less rich experience were it not for the groundbreaking research of the legions of scholars whose work over the previous three decades has fundamentally reshaped our knowledge of that event and its aftermath. But much had changed in those last thirty-plus years as well, not least of which was the abrupt end to the decades-long Cold War in 1989. And thus, if the Legacies conference was deeply indebted to post-Rhodesian research, it also presented an opportunity to take stock of that body of scholarship itself, to consider it as yet another of the Manhattan Project's diverse legacies.

1

Whether as a result of good timing or simply blind luck—or perhaps a bit of both—interest in the conference theme far surpassed the organizing committee's expectations. In one important respect, at least, that interest was a direct result of the former: When "Legacies" convened in March of 2017, work on the Foundation Document for the recently established Manhattan Project National Historical Park (or, somewhat curiously, "MAPR" in the acronym-obsessed patois of government-speak) was in full swing. Created by congressional legislation in November 2014, the park is atypical (if not entirely unique) within the country's national park system. Co-managed by the United States' Departments of Energy and Interior (DOE and DOI) and comprised of the three geographically dispersed "Secret Cities" sites—Hanford WA, Los Alamos NM, and Oak Ridge, TN—the Manhattan Project Park is the antithesis of its well-established and far more famous counterparts such as Yellowstone or Yosemite. In stark contrast to those shrines of Nature, where respite, retreat, reflection, and renewal are coins of the realm, Hanford, Los Alamos, and Oak Ridge are industrial sites most closely identified with toxic waste and massively expensive environmental cleanup. Little wonder, then, that the creation of a national park based on those sites stirred extensive criticism and controversy. Likewise, while issues of access, public safety, and security are concerns every national park site must address, the unusual circumstances at each of MAPR's three locations present logistical difficulties unthinkable in more traditional parks. Even interpretation poses a challenge. Indeed, in such a polarized and polarizing context, interpretation may be the most vexing challenge of all. Yet despite the controversy, or perhaps as a result of it, the creation of MAPR has proven a catalyst for renewed consideration of the Manhattan Project and its place in history. With its Foundation Document guiding the development and implementation of the park's interpretive framework, the National Park Service—in its appointed role as the "nation's storyteller"—has inserted itself squarely into the national dialogue. The roundtable discussion that brought the Legacies conference to a close (a lightly edited transcription concludes this collection) served as a forum of unsolicited advice for MAPR staff regarding future directions for that framework. While not an official part of the park's information-gathering process, the discussion among conference participants nevertheless captures a sense of the dialogue that guided its creation.

Fortunate timing was one obvious factor driving interest in the Legacies conference: the three Manhattan Project communities were deeply engaged at the time in the initial planning processes for the new park, and the conference offered an additional venue for facilitating those conversations. But the appeal of the conference theme went well beyond a merely local interest. Evidence of a broader enthusiasm was borne out in at least two ways. First, while the meeting was initially conceived along traditional academic lines, it quickly became apparent it was not to be constrained within those parameters. As planning proceeded, the list of those wanting to participate expanded to include working scientists, government employees, retired health physicists, downwinders, representatives from community groups, and impassioned lay people, as well as scholars from across the United States and at least six different countries, working in a host of different academic fields. As one might expect in drawing from such an extensive range of expertise and backgrounds, this diverse gathering resulted in a genuinely remarkable exchange of ideas and thus provided a more fertile and thought-provoking conference experience than its organizers could possibly have envisioned or dared to have hoped.

An even more persuasive (and gratifying) sign of the heightened interest in reevaluating the Manhattan Project and its legacies is displayed in the quality and breadth of the essays gathered here. Covering an imposing variety of topics, from print journalism, activism, nuclear testing, science, and education to health physics, environmental cleanup, and kitsch, these essays, taken as a whole, both deepen our understanding of familiar matters and illuminate historical corners and crevices unexplored by an earlier generation of scholars. But they also serve another equally valuable purpose. They remind us, as one of William Faulkner's characters famously remarks, that "the past is never dead. It's not even past."[1] Never quite securely in the historical rear-view mirror (however much we might wish it so), the past remains instead, Faulkner suggests, very much with us, active and alive, the warp and woof making up the fabric of our here and now. T. S. Eliot has described this historical unfolding as the "presentness of the past."[2] But as Eliot notes elsewhere, past and present—a present suffused with the past—are equally prologue to the future: "Time present and time past / Are both perhaps present in time future / And time future contained in time past."[3] In thinking about the essays that follow in this context of the interplay of past and present, current events appear almost supernaturally conjured as if to underscore Eliot's observation. Quite literally, roughly thirty miles away and on the very day these words are being written in the spring of 2018, a group of visitors is touring Hanford's B Reactor. Not an unusual event in itself, except that with this delegation is the first-ever *hibakusha*—the name given to the survivors of the atomic bombings of Hiroshima or Nagasaki—to visit the plutonium production facility.

Ironies abound. Without the admirable efforts of the B Reactor Museum Association (BRMA), B would have been demolished or cocooned just as all the other Hanford nuclear facilities have been or will soon be. Without B, there would have been little basis for a national park at all, since access to the Los Alamos and Oak Ridge sites remains almost entirely restricted. As noted above, B Reactor produced the plutonium slugs fueling the Fat Man bomb that devastated Nagasaki. The atomic bombings of Hiroshima and Nagasaki gave rise to a new type of war survivor, the *hibakusha* or, literally, "explosion-affected people." Now, just a little less than three-quarters of a century later, a historical chain-reaction was coming full circle: Largely because B Reactor had been saved, the Manhattan Project National Historical Park had been created. National parks in turn are sites of interpretation and, too often, of genuflection. On this day, however, a native Nagasakian and *hibakusha* was troubling a U.S.-centric view of the Manhattan Project into *reflection* by completing his historic journey from one ground-zero to another.

It would be all too easy to go on at length, teasing out the layers of irony and complexity this story entails. That job is for another place and time. Suffice to observe here only that iconic B bears equal witness to the extremes of our own humanity: the awe-inspiring capacities of human intellect and human folly; the human capacity to create and the human capacity to destroy; and the uniquely human capacity to harness nature for the advancement of technology and industry coupled with the uniquely human capacity to become enslaved by those very technologies themselves. No ghost could possibly summon a more appropriate image of the human predicament as we transitioned from one decade, one century, one millennium, to the next. So we will leave the supernatural aside. This is, after all, human history, the history of our own making. The only ghosts in this story are those filaments of the past that continue to shape, and sometimes disturb, our own time.

As the following essays attest, the Manhattan Project's afterlives animate the present in profuse and often urgent ways. To give just one arresting example: As Ian Graig recounts in his contribution, the Federation of Atomic Scientists (FAS) was first formed in 1945 by Manhattan Project scientists concerned about the ethical and responsible use of the new nuclear technologies they had developed. Part of the broader post-war movement of an increasingly politically engaged scientific community described by Graig, the FAS (rebranded as the Federation of American Scientists in 1946) has continued ever since in its mission to "strive to make the world a safer, more informed place."[4] Through its journal *Bulletin of the Atomic Scientists* and its associated website, the FAS has maintained and adjusted its famous Doomsday Clock.[5] In 2018, as the President of the United States and the Supreme Leader of North Korea (or "Deranged Maniac" and "Rocket Boy" as they endearingly referred to each other) escalated their verbal sparring, the Doomsday Clock edged perilously close to midnight. In a lengthy statement released through the *Bulletin* website, the FAS outlined in great detail the risks posed by nuclear weapons brought about by progressively more reckless and belligerent words and actions. "Because of the extraordinary danger of the current moment," the report announced, "the Science and Security Board today moves the minute hand of the Doomsday Clock 30 seconds closer to catastrophe. It is now two minutes to midnight—the closest the Clock has ever been to Doomsday, and as close as it was in 1953, at the height of the Cold War."[6] In a head-spinning turn of events, the two adversaries suddenly reversed course and agreed to face-to-face talks—the first-ever such meeting of a U.S. president and a North Korean head of state. Despite the president's subsequent declaration of the close bond forged between the two as a result of their meeting, the Doomsday Clock has not moved back. On the contrary, such unpredictable behaviors are one of the reasons cited by the report for the increased risks.

The example of the Doomsday Clock poised on the brink of midnight is, admittedly, an extreme one. Yet we inhabit a world characterized by extraordinary complexity, an intricately interconnected "global village" which is nevertheless beset by threats and challenges inconceivable to earlier generations. How, for example, are we to "dispose" of radioactive waste with a half-life of 24,000 years? In a world in which, even to this day, nuclear annihilation remains the push of a button or mishap away, even the most seemingly benign of afterlives—war-time censorship, say, or the fetishization of the past—have the potential to suddenly turn ominous and pressing. In such daunting circumstances there is solace to be found in the audacious example of the Manhattan Project scientists: Through the application of their rational faculties, they sought to comprehend, and then to solve, their own monumental challenges. The essays collected here share that same commitment to critical thinking in the service of advancing knowledge and understanding. Taken together, they provide grounds for optimism and chart a path of hope for the future.

Notes

1. William Faulkner, *Requiem for a Nun,* in *Novels: 1942-1954* (New York: The Library of America, 1994), 535.

2. *Selected Prose of T. S. Eliot,* ed. Frank Kermode (New York: Farrar, Straus, and Giroux, 1975), 38.

3. T. S. Eliot, "Four Quartets," in *Collected Poems, 1909-1962* (London: Faber and Faber, 1963), 189.

4. https://fas.org/.

5. https://thebulletin.org/.

6. https://cdn.thebulletin.org/sites/default/files/2018DoomsdayClockStatement.pdf.

TRUTH IS THE FIRST CASUALTY OF WAR

CHAPTER ONE

Atomic Legacies in Censored Print
Newspapers and the Meaning of Nuclear War

Hilary Dickerson

INTRODUCTION

In March 1946—three months after one hen in Hiroshima started to lay eggs again and one sterile woman became "grateful" to the atomic bomb for her unexpected pregnancy after the *Enola Gay*'s bombing of the city—thirty Americans volunteered their service as patriotic guinea pigs for the United States' atomic test at Bikini Atoll.[1]

Seventy years on, the tragedy, bravado, and sensationalism of these stories is complicated by the public venue in which they were circulated: the Japanese English-language dailies the *Mainichi* and the *Nippon Times*. At the time that these stories went to press, both newspapers were firmly under the aegis of General Douglas MacArthur and the American Occupation, as part of the charge to cultivate Christian democracy in occupied Japan. As Americans purposefully focused on the earth-shattering power of their newly minted weapon, Japanese *hibakusha* (atomic bomb victims) lived a far different reality, one that brought into stark relief the long-ranging human cost of war. Out of this post-war environment emerged attempts to memorialize Hiroshima and Nagasaki and to generate a memory of the war that opposed President Harry S. Truman's triumphant assertion of God's divine selection of America as the bomb's benefactor.

This essay examines journalistic accounts of atomic warfare. By returning to Japan's Occupation (1945–1952) and its occupied press, we can trace counter-narratives that emerged from Japan and the United States which questioned atomic warfare and America's identity as an

9

exceptional Christian democracy. In particular, we examine three stages of the atomic narrative that appeared in the English-language dailies: the days after the bombing of Hiroshima and Nagasaki; the transition from militarist censorship to Occupation censorship; and the early months of the Occupation itself. Newspaper articles provide a tour of international perspectives on Hiroshima and Nagasaki's widespread aftermath, allowing readers to trace the contested spaces in the American atomic narrative— long lionized for its termination of the war—more than seventy years later. From 1945–1946, even though press coverage about the atomic bomb was censored in various ways on both sides of the Pacific, the triumphalist rhetoric that would eventually prevail (as in, for example, the 1995 National Air and Space Museum's *Enola Gay* exhibit) had yet to take root. Rather, the English-language dailies published in Japan reveal the conflicted interpretations of Hiroshima and Nagasaki that ran in American and Japanese newspapers and subsequently complicated national portrayals of war and peace.

"MAN IS TOO FRAIL TO BE ENTRUSTED WITH SUCH POWER": THE ATOMIC BOMB BETWEEN CENSORS

In the days following August 6 and 9, 1945, after the United States bombed the city centers of Hiroshima and Nagasaki, Japanese English-language newspapers used the advent of atomic warfare to remind the United States (portrayed as the aggressor in the last weeks of August) of the vast gulf between its claims to righteousness and its actual behavior. Despite myriad constrictions to freedom of the press in the last days of the Asia-Pacific war, newspaper reports in Japan remain significant to this day for their early reactions to Hiroshima and Nagasaki that predated formal Occupation censorship. Articles condemned wartime policies that targeted civilians and created nuclear weapons, reminding Americans of their Christian duty. They also measured the destruction of Hiroshima and Nagasaki and, in doing so, chronicled the demise of the Japanese militarists' censorship of Japan's news. Additionally, they debated the peacetime roles of the only two cities leveled by atomic weapons.

The post-atomic bomb and pre-Occupation editions of the *Nippon Times* emphasized the international discomfort with the nuclear incineration of Hiroshima and Nagasaki. As with articles published earlier that August, the *Nippon Times* relied on wartime hostility as it depicted its enemy's and soon-to-be-occupier's devaluation of Japanese civilian life. Citing "a Reuters newscast received in Stockholm" on August 9, the newspaper informed its readers that Truman's atomic bomb "was bitterly criticized by the Vatican spokesman, who said that the news created a 'painful impression.'"[2] Although Pope Pius XII later deemed this comment "unauthorized," the *Nippon Times* leveraged international distaste for American foreign policy to contend, "It seems that the enemy is now intent on killing and wounding as many innocent people as possible due to his urgent desire to end the war speedily."[3] Subsequent articles revisited this theme of unjust war conduct, noting that the "atrocious action of the enemy" had disfigured Hiroshima's schoolchildren.[4] By August 18, the *Nippon Times* was using Domei News, the official news agency of imperial Japan, to narrate Hiroshima's horror through American witnesses. *Enola Gay* pilot Colonel Paul Tibbets "said it was difficult to believe 'what we did,'" while "Commander [William Sterling] Parsons of the United States Navy who participated in making the bomb and accompanied the bombing flight" called Hiroshima, "colossal and horrifying."[5] Coverage such as this—intended to condemn the United States' atomic bomb for foreign policy propaganda—placed Domei News in General MacArthur's crosshairs by the start of the Occupation in early September.[6]

In the week after Hirohito's August 15 radio broadcast announcing Japan's capitulation, a somber tone overtook the *Mainichi*'s reporting on the atomic bomb. Previously its coverage had been curtailed by the militarists' news censorship, leading to the August 11 claim that Nagasaki's "damage . . . is surmised to be extremely light." Now, however, two photographs of Nagasaki ran juxtaposed next to the editorial "A New Mentality." Not yet front-page news, Nagasaki's remains appeared grainy, portrayed as "Scorched Earth in A Moment," and described in the stark title "Houses 16 Kilometers Away Destroyed by Heat of Bomb." Citing Hirohito's surrender broadcast, "A New Mentality" explained the new duty of the Emperor's loyal subjects: "Bracing up our spirit amidst all sorrows, we should forge ahead for the construction of a new Japan."[7]

International journalists who visited Hiroshima in the early weeks after the city's near eradication emphasized prewar ties between the belligerents while contrasting the abject devastation with sights in war-torn

Europe, confirming Japan's sorrows. On September 4, an unnamed NBC correspondent detailed his trip to Hiroshima as a "scene which was gaunt, black and cruel," but one where the people were "not unfriendly." NBC's anonymous reporter met elderly men who explained in English that they had once lived in California and "even smiled when we said we were Americans."[8] An unnamed "British scribe," said a *Nippon Times* reprint of a Domei News release (published just days before MacArthur's virtual gutting of the news conglomerate) concluded that "few people in Britain, even those in the most badly bombed areas, could imagine the destruction caused by a single bomb" and warned that even "a month after the bomb was dropped, the stench of the dead is terrible, worse than those of the battlefields of Normandy."[9] Two days later, *Nippon Times* staff cited the reactions of some American reporters who, after their September 4 visit to Hiroshima, stated that the city's damage was "far more severe than anything they had seen in any European city."[10] Comparisons between European and Japanese suffering elucidated the unfathomable destruction endured in Hiroshima and Nagasaki and implied that Japan suffered far more than the Allies had during either the Blitz or the D-Day invasion. The timeline in which these accounts appeared in Japan's English-language newspapers marked the diminishing criticism of the Allies— vis-à-vis the atomic bomb—allowed by MacArthur in the following weeks.

Four days before Supreme Commander of the Allied Powers (SCAP) General Douglas MacArthur landed at Atsugi Air Base to begin Japan's occupation, the *Nippon Times* published "Atomic Bomb," an editorial by nom-de-plume Japonicus, who questioned the pairing of America's Christian democracy with the nuclear incineration of entire cities. Juxtaposing the United States' professed value system with its actual behavior, the anonymous author used the voices of American dissenters to reveal that even the United States' own Christian citizens did not accept the inherent evil of destroying Hiroshima or their government's justification for it:

> Religionists in America are reported to have condemned the measure at least in so far as it was employed against a populous city. The Americans, they say, should have tried it on a smaller Japanese town. The pronouncement is based on considerations of inhumanity, though not basically objecting to the employment of the bomb, a stand, we are afraid, rather ill-becoming persons preaching the Christian faith.

Japonicus also chronicled the ironic timing of the Nuremburg trials. Prominent Nazis faced their own war crimes and violations of international law on the heels of Hiroshima's destruction, an act "with malice aforethought and without any previous notice to destroy a large and populous town in defiance of international treaties and international morality." For Japonicus, this was evidence sufficient to "impress the world at large that there is one law for the conqueror and another law for the vanquished."[11]

In the weeks after August 9, newspaper articles documented the power of atomic weaponry and proved its destabilizing dangers not only for Japan but for all of humanity. Released the day before MacArthur and potential press censorship arrived, the *Mainichi's* article "Terrific Power Of Atom Bomb" recorded the civilian deaths, the loss of hospitals and schools, the "sort of derangement [sic]" that "occurs in the bodies of those walking around," and the death of even "worms and moles in the earth" at Hiroshima and Nagasaki. As the article's title warned in a prediction that both cities later struggled to overcome: "No Living Things Able to Exist in Areas For 70 Years." Within its scientifically detailed exposure of how victims suffered and died, based on their location inside or outside shelter and their distance from the bomb's hypocenter, the reporting embodied what later became part of the peacetime identities of Hiroshima and Nagasaki that, in their emphasis on the ways in which nuclear war threatened all humanity, frequently conflicted with American recollections of the bombings. The *Mainichi* suggested that the "ruined cities of Hiroshima and Nagasaki be left as they are as war monuments for ever [sic] so that the terrible nature of the atom bomb may be made known to all the races of the world." Reminding its readers about who had perpetrated both acts, the *Mainichi* noted that "radio announcements from the United States" had "revealed much of the horrible nature of the atom bomb," and pointed to what would become the driving force behind peace memorials created in Hiroshima and Nagasaki in the decades to come: "Thus it is keenly felt that its inhumanity should be condemned in the name of all mankind."[12]

The same edition of the *Mainichi* that predicted a peace identity for Hiroshima and Nagasaki also included statistics measuring Japan's wartime devastation that were unlikely to have seen print during the strict wartime censorship of the press. In the August 29 editorial "Reconstruction of

Destroyed Cities," an unidentified essayist cited an "official report" of the nation's wartime damage: 9.2 million Japanese "had lost their dwellings" "either by burning or by bombing" in attacks that killed 260,000 civilians—an estimate "feared to grow as investigation progresses"—and injured 420,000. Noting that these figures excluded military personnel and referred only to "damage caused to civilians," the author continued: "The pity of it all is that the war has been lost despite this appalling cost and the nation must bear it together with the further burdens to be imposed upon it by the victors." The editorial listed the physical, material, and psychological needs of the war's victims, recording for its audience the war's toll before the Occupation's arrival. Beyond the food shortages, loss of homes, suffering citizens living in bombed-out cities, and other tragedies that necessitated immediate attention from Japan's government, the anonymous author pinpointed physical ruins as symbolic of the nation's psyche: "The nation is now fed up with its unfortunate war experiences and is eager to be relieved of ruins that stir its bruised heart. The elimination of these eye-sores should be the starting point for a national renaissance as well as for relief of war victims." Even as the author argued against maintaining the sort of ruins that later became central to peace monuments in Hiroshima and Nagasaki, the reasoning in the editorial matched struggles in each city to move on with life, to have an identity more complex than victimhood, and to call the world's attention to the long-lasting devastation of war.[13]

As Occupation press censorship began, the *Mainichi* published two articles that marked the briefly freed space between old state censorship and new. "'1 Bomb Did All This,' Only Thought of Correspondent Visiting Hiroshima," an article by United Press war correspondent James McGlincy, said Hiroshima "once was a prosperous modern city where Japanese who returned from America liked to settle." McGlincy documented the smell of death that permeated the area and the toll of radiation sickness on the human body, as explained by Japanese physicians. The city's death, McGlincy wrote, "is indescribable and nobody in America can ever know what it is like unless he has seen it or—God forbid—unless it some day falls on America." Assisted by a Sacramento, California-born translator, McGlincy backed his assertion that "[i]n this city you can see in the eyes of the few Japanese picking through the ruins all the hate it is possible for

[a] human to muster" with his guide's own confirmation. When McGlincy asked, standing in rubble birthed by the United States, "How do people here feel about us? Do they hate us or do they think it is the fortune of war?" his translator replied, "They hate you."[14] McGlincy's reporting bridged the implementation of SCAP censorship, as his reference to the *hibakusha*'s hatred of Americans—and criticism of the atomic bomb and thus the United States—indicates. The *Mainichi*'s three-part series, "Force of the Atomic Bomb Great," started the week after Japan's formal September 2 surrender onboard the USS *Missouri*. While large portions of each day's installation covered historical aspects of the physics that led to the production of uranium and plutonium, to the Manhattan Project, and ultimately to the splitting of the atom, the final article concluded with international reactions to Hiroshima. Printed on September 13 in Japan, and running next to a photograph of Allied surgeons examining and treating one of Hiroshima's *hibakusha*, one installment of the "Force of Atomic Bomb Great" series cited a recently published *New York Times* article and its inclusion of secular and religious responses to atomic warfare. One anonymous American had written, "It is a stain upon our national life," and another suggested, "man is too frail to be entrusted with such power." Representatives for the Federal Council of Churches had opined, in rhetoric that prompted U.S. citizens to imagine their own imminent destruction rather than the death already occurring in Japan, "If we, a professedly Christian nation, feel morally free to use atomic energy in that way, men elsewhere will accept that verdict—the stage will be set for the sudden and final destruction of mankind." Cecil Hinshaw, president of the Quaker institution William Penn College, believed the "atomic bombing was a barbaric, inhuman type of warfare—its use unjustified." In an abrupt conclusion, the *Mainichi* turned to Nazi Hermann Goering "nervously awaiting trial as a war criminal." The once high-ranking Nazi general and head of the Luftwaffe had "heard the news [of the atomic bomb] in awe and wonderment" and then surmised: "Mighty accomplishment. I don't want anything to do with it. I am leaving this world."[15] A compliment on the ingenuity of American-led science by one of the architects of Nazi extermination policies was a denunciation of the United States' newest achievement, and it appeared in the transition between censorship by Japan's militarists and censorship by Japan's occupiers.

Demonstrating "A Right to a Place among Civilized Nations": America's Bomb in Occupied Japan's Print

Eight days after the surrender ceremony aboard the USS *Missouri* ended fifteen years of war in Asia, American Occupation authorities passed censorship laws that went into effect on September 10, 1945. These laws sharply curtailed what might have been the press's newfound freedom by forbidding printed or radio reports on "Allied troop movements which had not been officially released, false or deceptive criticism of Allied powers, and rumors."[16] MacArthur's General Headquarters (GHQ) had only to step into the propaganda and censorship apparatus Japan's government had long fostered.[17] By September 15, SCAP announced that Occupation censors would impose "100 per cent censorship of Japanese news" as, according to chief of censorship Colonel Donald Hoover, "Marshal MacArthur wants it understood that the Allied powers do not regard Japan as an equal in any way." Hoover stated, "Japan is a defeated enemy which has not yet demonstrated a right to a place among civilized nations."[18] Under these regulations, incendiary rhetoric like that levied against the United States in early August disappeared from Japanese English-language newspapers.[19]

SCAP's treatment of certain members of Japan's press during the Occupation's early days leaned toward retribution for their recently published articles and for their general tone throughout the war. Both the Domei News Service and the *Nippon Times* felt the sting of national defeat by mid-September. Employees of the Civil Censorship Detachment (CCD), relying on decoded intercepts of messages relayed between Japan's Foreign Minister Shigemitsu Mamoru and "Japanese legations in Berne, Stockholm, and Lisbon," discovered the close interactions between Domei and Japanese foreign policy. The decoded messages indicated the government's plans to use Domei to "make every effort to exploit the atomic bomb question in our propaganda."[20] This revelation resulted in Domei's censure by the CCD on September 14. As punishment for "disseminating untruthful news" and using coverage of Hiroshima and Nagasaki to "[disturb] the public tranquility," an offense specifically forbidden by the new press codes, Domei found itself confined to "pre-publication censorship" and lost one full day of publicizing news.[21] SCAP interrupted the *Nippon Times'* publication schedule for not properly turning over "a sensitive article" to Counter-Intelligence Chief Elliott R. Thorpe. In perhaps the best example of the Occupation's ability to simply step into the shoes left vacant by the militarists' highly bureaucratized control of the populace's state loyalty, MacArthur followed his advisors' suggestion and did not dismantle Domei News. Instead, the Supreme Commander used it to disseminate American propaganda. Thorpe later recalled: "[W]e were using Domei as the only effective means of communicating the Commander-in-Chief's will to the nearly 60 million people in the Empire."[22]

SCAP censorship, under the Civil Intelligence Section and its CCD, permeated "every aspect of public expression" and prompted the familiar trope of "self-censorship" by Japan's media.[23] SCAP enforced censorship parameters from September 1945 until September 1949, employing 6,000 Japanese who spoke English in its effort to craft the postwar peace by monitoring seventy newspapers with "*pre*publication censorship."[24] Under the September 1945 Press Code, editors had to submit proofs for clearance to SCAP censors, then remove any problematic material before receiving permission to publish.[25] SCAP guidelines shifted over time, often confounding Japanese attempts to identify the nuanced line between "fit to print" and incendiary. SCAP officials never informed the Japanese what constituted "unacceptable expression," leaving them to rely on two elements for guidance in their publications. One was the "very general press, radio, and film 'codes'" from the Occupation's start in 1945 that broadly restricted news that might, for example, "disturb the public tranquility" or present "false or destructive criticism of the Allied Powers."[26] The September 1945 Press Code restricted the presence of "editorial opinion" within newspapers in an effort to extricate propaganda from daily circulation.[27] The second was what John Dower calls "imagination shaped by experience," essentially an attempt by the Japanese to divine what would or would not be flagged for removal by SCAP censors. Classified and generated month by month, censorship guidelines for SCAP's CCD, such as one from June 1946, covered topics as wide ranging as criticism of any of the Allies as specified by country and by "general criticism," "Third World War Comments," "Overplaying Starvation," "Criticism of the United States," "Criticism of SCAP Writing the Constitution [including any reference whatsoever to SCAP's role]," and "Untrue Statements." Absent from this list was any specific reference to the atomic bomb.[28] The GHQ hoped not only to remove potential criticism of the Occupation itself, but also to exterminate the ideas that the Allies believed had sparked war and might undercut Japan's new democracy.[29]

Despite the newly imposed, ambiguous, and metamorphosing boundaries on press freedom, English-language newspapers frequently ran articles discussing victims of the atomic holocaust, death tolls, and destruction estimates. Such articles were often either based on or reprinted from news conglomerates such as the United Press (UP) and the Associated Press (AP) at the behest of MacArthur's infrastructure itself. SCAP's Civil Information and Education Section (CI&E) utilized its Information Division to encourage the dissemination of American culture by plying Japan's print media with articles from the United States. The presence of the victor's culture was not small in scale. The Information Division provided Japan's news agencies with "350 to 400" dispatches each month, generated by the former Allies, SCAP itself, or the UN.[30] SCAP closely monitored the organizations it used to spread the conqueror's culture amongst the conquered. Before going into circulation in Japan's home islands, some AP and UP articles first had to pass the censors' scrutiny.[31] Under the vigilant watch of Occupation censors, English-language newspapers referenced American unease with Hiroshima and Nagasaki. Prohibited from publishing pointed critiques of the nuclear obliteration of both cities—language that might "disturb the public tranquility"—newspapers still provided commentary on the United States' behavior. Two days after the first censorship laws went into effect, and three days before the Occupation garnered complete control over news published or aired in Japan, the *Nippon Times* article "American Experts See Atomic Bomb Victims" quoted Brigadier General T. F. Farrell. In an interview that followed his visit to Hiroshima, arranged by MacArthur's GHQ, Farrell stated, "personally I have realized that the damage is beyond expression. I presume that such a cruel war arm should not be used in the future."[32] In December 1945, the *Mainichi* included a poll conducted in the United States by the Opinion Research Center that "revealed that 54 per cent of the people in the United States believe manufacture of atomic bombs should be declared an international crime."[33] While it lacked the direct editorializing forbidden by the Press Code, the article still implied that something deemed an "international crime" in December had been equally problematic four months earlier.

The Japanese press's transparent use of the United States' own military representatives to contradict Truman's decision to crush Japan's Greater East Asia Co-Prosperity Sphere with the atomic bomb disappeared as SCAP increased its control over news consumed in Japan. No longer as pointed as the news/propaganda printed in August and early September, English-language newspapers from late 1945 on still frequently referenced the physical and structural agonies endured in Hiroshima and Nagasaki, demonstrating the shifting nature of censorship regulations. Although depictions of conditions in the bombed-out cities often relied on the aegis of the American press, reflecting MacArthur's desire to disseminate American culture within occupied territory, articles generated by the English-language dailies themselves addressed life after the atomic bomb. Focused on the slow regeneration of Hiroshima's infrastructure, the *Nippon Times'* "Hiroshima Recovering from A-Bomb Ravages" detailed population statistics from pre- and post-August 1945. Even with its optimistic observation that the "population which dwindled to 88,000 has witnessed a steady increase to some 149,000 as of March 20" the article added, "Hiroshima city in prewar days boasted a population of some 400,000," leaving its readers to ponder the fate of those now absent from the city along with those who, still residing there, lacked reliable access to food eight months after "Little Boy's" blast and burn.[34] Reprints carried with them the American "can do" attitude emphasized in the rebuilding of Japan, which molded the United States citizenry's memory of the post-war world and of the atomic flattening of two cities populated by noncombatants, but they also included statistics that measured the massive destruction caused by the bombs. UP correspondent Glenn Babb's "Nagasaki Is Center of World-Wide Search for Knowledge to Cope with Atom Bomb" appeared in the *Nippon Times* in December 1945 and labeled the city as "one of the world's two great laboratories holding the mysteries of the atomic bomb." Admitting that "Fat Boy" fell off-target when the B-29 *Bockscar* released it just after 11:00 am on August 9, Babb omitted mentioning the Urakami Cathedral district under Nagasaki's "Ground Zero."[35] The city's three Mitsubishi war plants did appear, as did casualty estimates: "30,000 to 40,000 persons perished almost instantaneously when the bomb loosed the sun's own energy." The "subsequent six weeks" witnessed "an equal number" of deaths and "[a] few are still dying—three or four weekly."[36] In April 1946, the *Mainichi* updated its readers on Hiroshima's "Fast Reconstruction": as 220 residents per day moved into a city "where the bomb demolished about 90 percent of the homes," they found 10,000 newly constructed homes in the rubble—the "fast" rebuilding highlighted in the UP's title—but they rode on only thirty tram cars for a population of 169,000, lacked school facilities, and waited for hundreds of telephones in the city to be replaced, while Japan itself endured a postwar food shortage.[37] Hiroshima's school children attended classes outside "in the open air, in the midst of rubble and without a single leafy tree to offer shade" that May; despite optimism about the city's rebirth, a material shortage stopped plans to rebuild primary schools, leaving school schedules at the mercy of the weather.[38] Censored as it was, the English-language press depicted the losses, fears, and hopes of postwar Japan, either through its own reporting or via article reprints from the United States.

In the Occupation's first year, targeted condemnation of the American decision to use the atomic bomb was absent from the discourse in the English-language newspapers. By 1946, reprints from the United States and articles written by the English-language dailies addressed themes of rebuilding, regeneration, forgiveness, and hope in local and international realms. As a *Nippon Times* dispatch cautiously indicated in January, within the bomb's power to damage life appeared its limited ability to improve life as well. Summarizing a recent *Asahi* report, Hiroshima "still lies in devastation" from the "historic atomic bomb attack," the article narrated, but signs of revitalization had started to emerge, from one hen starting to lay eggs after a six-month hiatus taken by "chickens in that town" to "[a]t least five of many atomic-affected female patients . . . 'return[ing] to normalcy physiologically.'" The abnormal held a modicum of hope: "[S]ome women are now grateful to 'Mr. Atomic Bomb' because they who had thought themselves sterile have become pregnant after the atomic bomb explosion." In measured language, "Hiroshima Hens Begin to Lay Eggs Again; Sterile, Aged Woman Grateful to A-Bomb" cautioned that "these 'beneficiaries'" had not been near the bomb's hypocenter. The Hiroshima Red Cross Hospital's investigation showed that "the atom explosion can have some favorable effect too, when its radioactivity is not so strong."[39] Other reports disproved concerns that Hiroshima and Nagasaki would remain radioactive for decades—and potentially uninhabitable for seventy years. Mrs. Takeno Nakashima, visiting Tokyo to see her son, U.S. Army Sergeant Henry Nakashima, confirmed for the UP in February 1946 that "vegetables and wheat planted in the atom bombed areas by return-ing residents are thriving."[40] Russell Brines, writing for the AP in May 1946, explained in the breezily titled "New Hiroshima City to Be Best in Orient," that Hiroshima's city officials had planned a five-year-long rebuilding schedule. Financed by the city and the national government, it would terminate with "a new city dedicated to 'international amity.'" As part of this international status, the city planned to preserve the "skel-etal dome of the former Museum of Industrial Art": labeled by Brines a "ghostly monument to the explosion," the dome remains the focal point of Hiroshima's Peace Memorial Museum and Peace Park today.

Brines' analysis of Hiroshima's rebirth reflected the United States' insistence that a new and better Japan would emerge from American-led peace. It also returned readers to those in-between weeks of August and early September 1945, when the English-language dailies first advocated peace legacies for both cities. Brines compared atomic-bombed Hiroshima and its resurgent green grass with the fire-bombed capitol, and posited that the former was "neater, more alert and better fed than Tokyo." Although hopeful of symbolizing "international amity," Hiroshima's Mayor Kihara had abandoned his earlier request that Americans donate funds to the city's rebuilding.[41] Depicted through the Occupier's lens, the Hiroshima obliterated by the United States' atomic bomb was now more advanced than the heart of Imperial Japan, Tokyo, which had been severely scorched during the war by traditional mass bombing.

Some articles sent to press in Japanese newspapers mitigated American responsibility for the atomic bombings while simultaneously voicing the horrors lived in Hiroshima and Nagasaki even months after August 1945. Two Westerners familiar with Japan concluded in 1946 that the *hibakusha* were not angry at the United States for incinerating their cities, contrary to earlier international reports of palpable dislike for Americans in Hiro-shima, during the transition from militarist to Occupation censorship. Dr. Martin Hall had taught in Japan for twenty-three years and returned to the country as part of the Atomic Bomb Survey in 1946. Following one-month stays in Hiroshima and Nagasaki, Hall spoke at Doshisha University in Kyoto, where he had once taught English. "The atomic bomb sufferers of Hiroshima and Nagasaki have no resentment against the American people," said Martin Hall in the *Mainichi*'s summary of the meeting. "[T]hey only hope to rise together with the United States to create new peaceful world." Hall believed that the "sad" existence endured in Hiroshima and Nagasaki would "someday bloom out into a great monument of peace."[42]

Printed directly to the left of a front-page article lauding the "Historic Bikini Test" of July 1, 1946, the *Mainichi* piece "Japanese Not Angered by Atom Bomb Attack, Catholic Priest Avers" cited German Catholic Father John A. Siemes' experience in Hiroshima after the "historic blast" nearly a year earlier. A philosophy professor at Tokyo Catholic University by 1946, Siemes spoke to "the attitude of the populace when thousands of dead lay strewn in heaping wreckage and thousands more cried for help amid ruined buildings." He remembered: "None of us in those days heard a single outburst against the Americans on the part of the Japanese nor was there any evidence of a vengeful spirit. The Japanese suffered this terrible blow as part of the fortunes of war—something to be borne without complaint."[43] Siemes' abstract invocation of the "fortunes of war" as the cause of the attack, rather than the United States, paralleled American depictions of the bombing as a natural disaster and repudiated earlier reports that emphasized Hiroshima citizens' hatred of America. By narrating the *hibakushas'* lack of rage against the United States, the clergy-man confirmed the Japanese militarists' culpability for the war's end and removed any guilt associated with Truman's decision to flatten a civilian target. The article's reprint in the *Nippon Times* is thus unsurprising, as it solidified a version of Hiroshima acceptable to SCAP authorities in the very newspaper once used by Domei News as a vehicle for atomic narratives that generated anti-American sentiment.

These articles in Japan's English-language dailies reflected the United States' fascination with the power of its atomic creation, visible both in the fear that a nuclear bomb might detonate above American cities and in the hope that atomic energy would improve the world—even in the cities it reduced to ash. They illustrated MacArthur's intent to disseminate American culture within Occupied Japan but may also indicate Japanese agency in crafting atomic narratives. In a sensationalist turn in 1946, the *Mainichi* published a series of UP stories portraying conflicted American responses to nuclear warfare. In March, as the United States prepared for A-bomb detonations in the South Pacific, "[t]hirty persons including a parachute jumper, a self-professed alcholic [sic], several war veterans and a woman have volunteered to serve as human 'guinea pigs' in the atomic bomb tests" at Bikini Atoll. Although their backgrounds and ages were quite varied—ranging from a 72-year-old man who regretted that his age had limited his participation in World War II to "sit[ting] tight and pay[ing] taxes and buy[ing] war bonds," to an 18-year-old who had spent time in "serious consideration" before contacting the government to offer his participation—they nonetheless voiced devotion and patriotic fervor for America's post-war supremacy. And most of the thirty offered their lives for free: "Only two of them thought they (meaning next of kin) should be paid. One estimated that $50,000 would be 'about right.'" One volunteer explained, "Fifty per cent of our population are alcoholics and as an alcoholic I offer myself as a guinea pig. P.S. I bet you a thousand dollars I live through it," while another who had volunteered with a friend wrote, "We felt it our duty, not only to science but to the world." The volunteers celebrated the genius of American-led science and the duty to sacrifice for the purported good of humanity and commemorated yet another iteration of the "Good War's" national triumph and prosperity; they also marked the distance between the horrors lived and died at Hiroshima and Nagasaki and the American public's understanding of what had occurred in both cities. Thanking the "guinea pigs" for their "courageous offer," the government turned down their requests, irradiating only "goats, sheep, pigs and white rats" at Bikini Atoll and reassuring international observers "that no human beings will be intentionally submitted to the explosive tests."[44]

An October UP article in the *Mainichi* covering William Kennan, a "New Yorker" who was "In [an] Awful Hurry to Avoid A-Bomb" and moved his family to Montana to "be safe from World War III if it comes," countered the unbridled enthusiasm of the atomic volunteers. Kennan had "been rushing to get his private affairs in order since the atom bombing in Hiroshima."[45] Citing American apprehension of atomic warfare, the *Mainichi*'s reprint implied, without raising the ire of the Occupation censors, that the bomb's legacy included discomfort and fear even for the victors, and it did so while referencing World War III—a topic forbidden by SCAP's press code only four months earlier.

CONCLUSION

Japan's journalists and editors were not alone in questioning, constrained though it was, the American victory narrative. Subjected first to the push and pull of Japan's regulations and then to that of Occupation censorship, Japan's English-language newspapers published during the end of war and the start of peace illuminated attempts to subvert American claims of righteous victory. The transformation of perspective and tone evident in the *Nippon Times* and the *Mainichi* embodied the public interplay between the victors and the vanquished regarding portrayals of the war's end. While the United States held on to its sanctified position as a Christian democracy—indeed, disseminated its cultural values widely via Japan's Occupied press—the stories publicized by the English-language dailies marked the nuances of Hiroshima and Nagasaki. More than seven decades after that devastating atomic August, these reports complicate national narratives of the Asia-Pacific War. Once visible on both sides of the Pacific, these stories can again return us to the contested spaces of war and peace, defeat and triumph, victim and victor.

Notes

1. The author wishes to thank Noriko Kawamura and International Christian University for the Center of Excellence Fellowship that made this research possible.

2. "New-Type Bombs Used in Raid on Hiroshima," *Nippon Times*, August 9, 1945, 1.

3. "Force of Atomic Bomb Great: All Possible Steps Taken to Assure Workers' Safety; People Voice Fear of Discovery In 'Hour of Victory,'" *Mainichi*, September 13, 1945, 2; "New-Type Bombs Used in Raid on Hiroshima," *Nippon Times*, August 9, 1945, 1.

4. "Eyewitness Depicts Effects of New Bomb Used by the Enemy in Raiding Hiroshima," *Nippon Times*, August 11, 1945, 3.

5. "Effect of Atomic Bomb Described by B-29 Pilot," *Nippon Times*, August 18, 1945, 1.

6. For more information on MacArthur's punishment of Domei News, see Eiji Takemae, *The Allied Occupation of Japan* (New York: Continuum, 2002), 385. Translated and adapted from the Japanese by Robert Ricketts and Sebastian Swann. Originally titled *Inside GHQ: The Allied Occupation and Its Legacy* (Tokyo: Iwashi Shinsho, 1983).

7. "Hit by Atom Bomb," *Mainichi*, August 22, 1945, 2; "On Nagasaki City," *Mainichi*, August 11, 1945, 2; "Editorial: A New Mentality," *Mainichi*, August 22, 1945, 2.

8. "Grim Tragedy of Atomic Bomb Vividly Told by NBC Scribe Who Visited Hiroshima," *Nippon Times*, September 8, 1945, 3. The article includes injury and death statistics for Hiroshima, estimating that 220,000 of the city's 350,000 residents had been hurt in the bombings and stating that 68,000 had been confirmed dead as of September 1, 1945.

9. Domei News, "British Scribe Shocked by Hiroshima Bombing," *Nippon Times*, September 6, 1945, 1.

10. "More U.S. Newsmen See Remains of Hiroshima," *Nippon Times*, September 8, 1945, 2.

11. Japonicus, "Atomic Bomb," *Nippon Times*, August 26, 1945, 4.

12. "Terrific Power of Atom Bomb: Hiroshima, Nagasaki Virtually Blown Off Face of Earth; No Living Things Able to Exist in Areas For 70 Years," *Mainichi*, August 29, 1945, 2.

13. "Editorial: Reconstruction of Destroyed Cities," *Mainichi*, August 29, 1945, 1.

14. James McGlincy, "'1 Bomb Did All This,' Only Thought of Correspondent Visiting Hiroshima," *Mainichi*, September 12, 1945, 3.

15. "Force of Atomic Bomb Great: All Possible Steps Taken to Assure Workers' Safety; People Voice Fear of Discovery In 'Hour of Victory,'" *Mainichi*, September 13, 1945, 2.

16. "Censorship of News Broadcasts Starts," *Mainichi*, September 13, 1945, 1. For a survey of Imperial Japan's use of wartime censorship, see Samuel Hideo Yamashita, *Leaves from an Autumn of Emergencies: Selections from the Wartime Diaries of Ordinary Japanese* (Honolulu: University of Hawaii Press, 2005).

17. Takemae, *The Allied Occupation of Japan*, 384.

18. "100 Per Cent Censorship of News of Japanese Origin to Be Effected," *Mainichi*, September 18, 1945, 1.

19. See William de Lange, *A History of Japanese Journalism: Japan's Press Club as the Last Obstacle to a Mature Press* (Richmond, UK: Curzon Press Ltd, 1998). The United States had been planning its approach toward defeated Japan's newspapers since 1944 and the creation of the State-War-Navy Coordinating Committee (166–167).

20. Takemae, *The Allied Occupation of Japan*, 385.

21. Ibid.

22. Elliott R. Thorpe, *East Wind, Rain* (Boston: Gambit Inc., 1969), 190–191. Quoted in Eiji Takemae, *The Allied Occupation of Japan*, 384.

23. John W. Dower, *Embracing Defeat: Japan in the Wake of World War II* (New York: W. W. Norton & Company, 2000), 405.

24. Ibid., 407.

25. de Lange, *A History of Japanese Journalism*, 169.

26. Dower, *Embracing Defeat*, 410.

27. de Lange, *A History of Japanese Journalism*, 168.

28. Dower, *Embracing Defeat*, 411. As Takemae discusses in *The Allied Occupation of Japan*, the dangers posed by SCAP regulations over print media were often economic, particularly if a work fell victim to the censor's blue pen *after* publication or if the time required to clear the censors prevented a paper from publishing an article when it was still news (410, 429). SCAP's use of economics to control the press extended to its distribution of the scant post-war paper supply; newspapers currently in favor received more of the paper ration than those that had raised the ire of Occupation authorities (432–433).

29. As Takemae concludes in *The Allied Occupation of Japan*, newspapers were not the only vehicle used to spread notions of Japan's own war guilt; "[m]agazines, documentary films, newsreels, and books" surveyed wartime atrocities as well (396).

30. Takemae, *The Allied Occupation of Japan*, 396. The Information Division did not stop simply with providing the Japanese with access to "American culture and values" in the daily newspapers. The Information Division's CI&E Information Centers, built throughout Japan, hired "friendly American librarians" to catalogue "between 5,000 and 10,000 volumes and some 400 periodicals" along with records, films, and lecture series that highlighted the U.S.'s cultural identity. Approximately "2 million Japanese" visited these centers (396).

31. Dower, *Embracing Defeat*, 406.

32. "American Experts See Atomic Bomb Victims," *Nippon Times*, September 12, 1945, 3.

33. United Press, "Poll on Atom Bombs," *Mainichi*, December 7, 1945, 1. The next sentence of the two-sentence blurb adds, "Thirty-one per cent of the people polled said they favored a World Police equipped with the bomb."

34. "Hiroshima Recovering from A-Bomb Ravages," *Nippon Times*, April 17, 1946, 3.

35. "Fat Man's" plutonium-fueled blast took the lives of 8,500 of Urakami's 12,000 parishioners. The destruction—tangible in the loss of the church building and thousands of Catholics and the spiritual grief described by Takashi Nagai—inflicted against Nagasaki on August 9 lasted throughout subsequent decades and formulated a new identity for Urakami's Catholics. Out of Urakami's decades-long push toward restoration emerged a consciousness that called on victims, as moral witnesses, to symbolize the lasting horrors of war and to transform the present. This identity destabilized the American narrative of the war's sanctified end. See *The Restoration of Urakami Cathedral, a Commemorative Album*, edited by Shohachi Hamaguchi and Hisayuki Mizuura and translated by Edward Hattrick (Nagasaki: Seibo no Kishi Publishers, 1981); Takashi Nagai, *We of Nagasaki: The Story of Survivors in an Atomic Wasteland*, translated by Ichiro Shirato and Herbert B. L. Silverman (London: Victor Gollancz, LTD, 1951).

36. Glenn Babb/United Press, "Nagasaki Is Center of World-Wide Search for Knowledge to Cope with Atom Bomb," *Nippon Times*, December 2, 1945, 3.

37. United Press, "Fast Reconstruction Seen in Hiroshima," *Mainichi*, April 18, 1946, 2; United Press, "Little Hope Is Seen About Food Shortage," *Mainichi*, April 18, 1946, 2.

38. "Education of Children Must Go On—Hiroshima Holds Classes in Open Air," *Mainichi*, May 18, 1946, 2.

39. "Hiroshima Hens Begin to Lay Eggs Again; Sterile, Aged Woman Grateful to A-Bomb," *Nippon Times*, January 25, 1946, 3.

40. United Press, "No More Atomic Bomb Effects Seen in Hiroshima And Nagasaki Cities," *Mainichi*, February 13, 1946, 1.

41. Russell Brines/Associated Press, "New Hiroshima City to Be Best in Orient," *Nippon Times*, May 28, 1946, 2.

42. "Hiroshima, Nagasaki Experiences are Told," *Mainichi*, May 19, 1946, 2.

43. "Japanese Not Angered by Atom Bomb Attack Catholic Priest Avers," *Nippon Times*, July 2, 1946, 1. INS. Washington, D.C.

44. United Press, "30 Offer to Be Human Guinea Pigs in A-Bomb Test; Get Courteous 'No,'" *Mainichi*, March 29, 1946, 2; United Press, "Indian Offers Himself as 'Human Guinea Pig," *Mainichi*, July 1, 1946, 1.

45. United Press, "New Yorker Is in Awful Hurry to Avoid A-Bomb," *Mainichi*, October 4, 1946, 1.

CHAPTER TWO

Borrowed Chronicles

William L. "Atomic Bill" Laurence and the Reports of a Hiroshima Survivor

Susan E. Swanberg

INTRODUCTION

William L. Laurence, *New York Times* (*Times*) science journalist from 1930 until he retired in 1964, wrote about the birth of the Atomic Age.[1] "Atomic Bill" Laurence, who was embedded with the United States War Department for nearly four months in 1945 writing press releases and articles about the Manhattan Project, returned to the *Times* after the war ended. In 1946 he won a Pulitzer Prize for eleven of his articles about the atomic bomb—all published in the *Times*—including an eyewitness account of the bombing of Nagasaki.[2]

Early in the twenty-first century, Amy and David Goodman, Beverley Deepe Keever, and others accused Laurence of being a War Department propagandist and cheerleader whose dual roles as a "special consultant" to the Manhattan Project and a *Times* science journalist raised serious ethical concerns.[3] He was also accused of complicity in U.S. efforts to conceal the impact of radiation sickness on survivors of the Hiroshima bombing. In at least two *Times* articles from September of 1945, Laurence contended that Japanese claims of the lingering effects of radioactivity were merely propaganda—an assertion contradicted by the earlier reporting of Wilfred Burchett, the first journalist to visit Hiroshima after the bomb was dropped.[4] Failing to write objectively about the Manhattan Project and the Hiroshima aftermath was not the only questionable behavior in which Laurence engaged. This essay reveals his previously unrecognized

29

appropriation of the writings of another author, and places Laurence's plagiarism in its historical, legal, and ethical contexts.

BORROWED CHRONICLES

As an important disseminator, if not a creator, of a narrative that persists to this day—that the use of the atomic bomb against Japan was justified—Laurence is worth revisiting through his writings about the bomb and the source materials upon which he relied.[5] In 1946 and 1959 respectively, he published two books about the atomic bomb, *Dawn Over Zero: The Story of the Atomic Bomb*, and *Men and Atoms: The Discovery, the Uses, and the Future of Atomic Energy*. In both books, Laurence borrowed extensively from the writings of Hiroshima survivor Father John A. Siemes, S.J., failing to quote, cite, and acknowledge Siemes' work adequately.[6] Laurence also altered sections of Siemes' text, changing a predominantly first-person memoir to a third-person narrative—a transformation that evidenced Laurence's active, conscious engagement with Siemes' material.[7]

Father Siemes, a German Jesuit priest and scholar who taught philosophy at a Catholic university in Tokyo, fled the U.S. firebombing of that city with a group of his students, taking shelter at a Jesuit novitiate on the outskirts of Hiroshima.[8] Siemes was living at the novitiate along with several of his Jesuit colleagues when the United States dropped the atomic bomb on Hiroshima. Siemes wrote a taut account of the bombing and its aftermath,[9] and in 1945 and 1946, he also appeared in two War Department propaganda films (*The Atom Strikes!* and *A Tale of Two Cities*), recounting portions of his eyewitness report. In the second of the two films, the U.S. War Department omitted Siemes' discussion of the morality of using an atomic bomb.[10]

Versions of Siemes' written report from Hiroshima appeared in a number of publications, with some variation in length, language, and punctuation among the versions.[11] His original eyewitness report is likely the account mentioned by Averill A. Liebow in his book, *Encounter with Disaster: A Medical Diary of Hiroshima, 1945*.[12] Liebow, an Austrian-born, Yale physician, worked in Japan from September 1945 to January 1946 as a member of the Joint Commission for the Investigation of the Effects of the Atomic Bomb in Japan, tasked with investigating the effects of the atomic bomb in Japan.[13] In a diary entry dated September 27, 1945,

Liebow recounted that U.S. Colonel Stafford L. Warren, chief radiologist for the Manhattan Project, asked Liebow to translate Siemes' statement from the German.[14] Liebow dictated the translation to a sergeant from "General Farrell's Manhattan District Group" [sic], and noted in a comment added to his diary before it was published in 1965 that "Siemes' report became a major source of material for John Hersey's masterful *Hiroshima*, and it was published in *full* [emphasis provided], in [his] impromptu translation, in *The Saturday Review* [sic] several years later."[15]

Liebow's statement was not entirely accurate. The Liebow translation comprised eight single-spaced pages of text. The version of Siemes' report appearing in the May 11, 1946, issue of *The Saturday Review of Literature* (SRL) was not the complete report. Several omissions involved a few words or phrases in some places, but in one case, three paragraphs of material were missing from a single page. Although many descriptions of casualties resulting from the bombing remained in the SRL article, much of the material missing from this version of Siemes' report contained his original eye-witness reporting of bomb-related casualties.[16]

As will be reported more fully below, Laurence used text from one or more versions of Siemes' reports in his book, *Dawn Over Zero*. While he listed Siemes' name in the book's index and frequently mentioned Siemes' name in the text, he failed to adequately distinguish quoted material from that which he paraphrased. Laurence also transformed much of the first-person account into a third-person narrative, and did not provide adequate acknowledgment or a definitive citation for the Siemes material.[17]

Laurence repeated his appropriation of Siemes' account in a second book, *Men and Atoms*. Again, he transformed much of Siemes' eyewitness report from a first-person voice to third-person and failed to provide a definitive citation for or acknowledgment of Siemes' work.[18] In both books, Laurence also recycled his own writings. Both contain excerpts from eleven of his *New York Times* stories—in some instances repeating nearly word-for-word what he had previously written, as in his account of the bombing of Nagasaki.[19]

EXTENT OF THE BORROWING

In *Dawn Over Zero*, Laurence introduced the Siemes material with biographical information about Father Siemes and a reference to a version of the Siemes' report published in *Jesuit Missions*, but no citation.[20] In the first edition published by Knopf, Laurence acknowledged his gratitude to a number of military figures, including General Leslie R. Groves and Colonel Stafford L. Warren, the officer for whom Dr. Liebow translated Siemes' eyewitness report. Laurence also acknowledged his gratitude for the help provided by several scientists. Among the acknowledgments is an ambiguously worded statement thanking Arthur Hays Sulzberger, then president and publisher of the *Times*, and Edwin L. James, the *Times* managing editor, for "their many kindnesses, including permission to reprint some of the material in this book."[21] Whether the wording of this acknowledgement related to retrospective or prospective use of material in *Dawn Over Zero* was unclear.

In another version of his experiences and insights into the atomic age— *Men and Atoms: The Discovery, the Uses, and the Future of Atomic Energy,*[22] Laurence again borrowed extensively from Father Siemes' Hiroshima report, manipulating the text in a number of respects. In addition to transforming the text from a first- to third-person account, he frequently added text to Siemes' account without clarifying which material was Siemes' and which was his own.[23] Moreover, *Men and Atoms* (published by Simon & Schuster in 1959) had no index, and contained no notations specifically authorizing reuse of previously published *Times* materials. There were no specific references to Father Siemes or any version of Siemes' Hiroshima report in the acknowledgments or permissions. Omitting the authorship claims of others, Laurence nevertheless asserted the publisher's—and his own—exclusive rights to reproduce material in the book: "ALL RIGHTS RESERVED," the copyright notice proclaimed, "INCLUDING THE RIGHT OF REPRODUCTION IN WHOLE OR IN PART IN ANY FORM COPYRIGHT ©1946, 1959 BY WILLIAM L. LAURENCE PUBLISHED BY SIMON AND SCHUSTER, INC. ROCKEFELLER CENTER, 630 FIFTH AVENUE NEW YORK 20 N.Y."[24]

HISTORICAL CONTEXT: WILLIAM LEONARD LAURENCE AS PLAYWRIGHT, TRANSLATOR, JOURNALIST, AND AUTHOR

During his career at the *Times*, Laurence won two Pulitzer prizes for his reporting. He wrote extensively about matters related to basic science, health, and medicine, but "[h]is biggest exclusive was the dawn of the nuclear age," noted his obituary.[25] Laurence is less well known for his early, mostly unsuccessful literary efforts, which likely left an indelible mark on his concept of authorship. Leib Wolf Siew, as he was named at birth, spent his first fifteen or sixteen years in a small Lithuanian village.[26] As a youth, he contributed news items to Hebrew newspapers published in St. Petersburg, but later maintained that this was "just a sideline."[27] After emigrating to the United States and serving in World War I as a member of the U.S. Signal Corp, he attended Harvard and Boston University—studying philosophy, drama, and law—and had a short-lived career translating and adapting Russian plays.[28] According to his obituary, Laurence adapted three plays from the Russian.[29] In 1930, he translated Maxim Gorki's play, *The Lower Depths*, peppering the masterpiece with American slang.[30] Laurence's version of Gorki's play, titled *At the Bottom*, was presented at the Waldorf Theater in Manhattan that same year. The adaptation was not favorably reviewed;[31] nor was his theatrical adaptation of Leonid Andreyev's short story "Thought," an effort that was panned in the May 11, 1931, issue of *Time*.[32]

As his career as a playwright was stalling, Laurence turned to journalism, where he met with considerable success despite his lack of formal training. In 1926, after Laurence dethroned *New York World* editor Herbert Bayard Swope in the parlor game, "Ask Me Another," Swope invited Laurence to the newspaper's office for an interview.[33] According to Laurence's account, he was hired as a reporter on the spot.[34] In a 1956 interview, Laurence explained that, while at the *World*, he became "ipso facto the man to report any scientific event that took place in the city."[35] In 1930 the *Times* offered Laurence a job as a science specialist.[36] At first Laurence balked at the job, but when the *World* refused to give him a raise, Laurence accepted the *Times'* offer.[37] While at the *Times*, Laurence developed his signature modus operandi, covering scientific conferences where he could hobnob with scientists, and was soon hooked on writing about science.[38] In his capacity as a reporter for the *Times*, Laurence networked with top scientists and learned about nuclear fission and its implications.[39] Throughout his career, Laurence met and wrote about Neils Bohr, Enrico Fermi, Albert Einstein, Robert Oppenheimer, and other illustrious nuclear scientists of the mid-twentieth century.[40] In 1937, Laurence and four other science journalists (with whom Laurence later formed the National Association of Science Writers) won a Pulitzer Prize for their coverage of the tercentenary of Harvard University.[41]

Laurence had been at the *Times* for fifteen years, writing primarily about science (including stories about nuclear physics and atomic energy) when he was invited to work for the War Department's Manhattan Engineer District (MED), known colloquially as the Manhattan Project. Laurence was personally recruited by General Leslie Groves, leader of the MED, to write about "the intricacies of the atomic bomb's operating principles in laymen's language."[42] The ostensible reasons for choosing Laurence were described in the August 7, 1945, *Times* article. "The [War] [D]epartment's choice of Mr. Laurence was a natural one since he 'discovered' for newspaper readers the method by which atomic energy was released by uranium fission as long ago as May 1940," its author noted.[43] "In an article published by THE TIMES at that time Mr. Laurence told its readers how the new material, U-235, was the most tremendous source of power known on earth."[44] On July 16, 1945, Laurence observed the Trinity test—the first test of an atomic bomb—at Alamogordo, New Mexico.[45] Laurence's dramatic interpretation of this experience can be heard, narrated by Laurence himself, on Fred Friendly's 1950 radio program, "The Quick and the Dead."[46]

The eleven stories that garnered Laurence the 1946 Pulitzer Prize for reporting were based on his experiences while with the MED. Laurence also wrote press releases for the War Department, drafted a speech for President Truman about the bombing of Japan, traveled to Manhattan Project installations around the United States, and, in August of 1945, observed the bombing of Nagasaki from the cockpit of an instrument plane accompanying Bockscar, the B-29 that dropped a plutonium bomb on the Japanese city.[47] Laurence's description of the bombing mission was originally written as a War Department press release to be provided to editors of U.S. newspapers with the notation that the story could be published "with or without the use of Mr. Laurence's name."[48] After the Hiroshima and Nagasaki bombings, a number of Laurence's stories were rewritten by other journalists and published without Laurence's byline, a fact that irked him.[49] When his stories were eventually published in the *Times* (with Laurence's byline) they were accompanied by statements acknowledging Laurence's special role with the Manhattan Project.

Laurence's unique if often conflicted relationships with both the *Times* and the Manhattan Project foreground essential and timely questions regarding journalistic ethics and the effect journalistic credibility, or lack thereof, can have on the public understanding of science. Laurence's earlier ethical lapses—the allegations of propagandizing and misrepresentations of critical facts—have not been forgotten, and in fact were revived in the mid-2000s with Amy and David Goodman's (unsuccessful) campaign to revoke Laurence's 1947 Pulitzer prize.[50] Together such ethical failures as those Laurence committed beg the question of the extent to which they helped lay the foundation for later failures of science journalism and the erosion of public trust in science (as with the autism/vaccine, GMO, and global climate change controversies). We may never be able to determine the magnitude of Laurence's influence, but we do know that repetition of inaccurate messages, including messages about science, does increase belief in those messages.[51]

AUTHORSHIP: LEGAL AND ETHICAL CONSIDERATIONS

Lisa Ede, citing Elizabeth L. Eisenstein's book *The Printing Press as an Agent of Change*, argues that the concept of authorship is a relatively recent invention, spurred by the development of the printing press. Economic factors that influenced the business of printing, Ede writes, "played a strong role in necessitating copyright laws, the ultimate expression of our belief that writers . . . literally own their texts."[52] Authorship is protected by both legal and ethical sanctions. The authority for U.S. copyright sanctions against unauthorized appropriation of an author's work is derived from the U.S. Constitution, which granted to Congress the power to promote and regulate the writings and discoveries of writers and inventors.[53] The scope of this authority is embodied in copyright laws passed by Congress and interpreted by the U.S. federal court system.[54]

The United States Copyright Act of 1790 was the first in a series of federal statutes written with the goal of providing authors (initially only those authors who were citizens or residents of the United States) exclusive rights to their writings for a limited term.[55] Although the United States joined the Universal Copyright Convention in 1955, which provided some protection to foreign authors, the more extensive protections afforded by the Berne Convention, an international copyright treaty which protected foreign authors from copyright infringement, were not available until the United States joined the convention in 1989.[56] With a few exceptions, government documents and publications have never had copyright protection.[57]

Ethical sanctions for appropriating an author's work vary, depending on the context. In academia, journalism, and other professions plagiarism is regarded as a serious transgression that can lead to loss of reputation and termination of employment.[58] Forms of plagiarism include outright copying and patchwriting—a form of inept paraphrasing—without proper citation and referencing.[59] Describing the differences between plagiarism and copyright infringement, Roger Billings argues that "the word 'plagiarism' is often used interchangeably with 'copyright infringement,' but the two terms are not synonymous." "Instead," Billings contends, "plagiarism is a state-based tort that has survived as a remnant of the nearly extinct field of common law copyright. Common law copyright is simply that which is not preempted by the [Copyright] Act…Plagiarism is the borrowing of someone else's work without attribution."[60]

In *The Little Book of Plagiarism*, eminent legal scholar and judge Richard A. Posner argues that "not all plagiarism is copyright infringement and not all copyright infringement is plagiarism."[61] Copyright law, according to Posner, protects only the form in which ideas or facts are presented, not the ideas or facts themselves.[62] The fair use doctrine, which permits some word-for-word use of copyrighted material in delineated instances, does not relieve a writer of the responsibility to place quoted copyrighted material within quotation marks or block quote format and to acknowledge the source. The essence of plagiarism, writes Posner, is concealment that misleads the reader such that the reader relies, to his or her detriment, on a representation of authorship by the writer.[63] Similarly, clarity and confusion are at the crux of John Higham's "Habits of the Cloth and Standards of the Academy." For Higham, plagiarism "obstructs the testing and validation of knowledge by hiding its true sources…" while the plagiarist "violates the code of a truth-seeking community by appropriating for himself the distinctive form in which someone else has tried to make a contribution." "In both respects," Higham concludes, "plagiarism sows confusion and weakens morale in the community it strikes."[64]

Piracy in Journalism during the Nineteenth and Twentieth Centuries

Journalism's attitude toward plagiarism has evolved considerably during the last 150 years. It is a well-documented fact that newspapers in the late nineteenth and early twentieth centuries sometimes stole and rewrote one another's stories.[65] A notorious example of journalistic piracy involved the *New York World*, the paper where Laurence began his journalism career in 1926. In 1898 the *World*—run at the time by Joseph Pulitzer—rewrote, published, and copyrighted a fake news story about the alleged death of "Austrian artillerist" Colonel Reflipe W. Thenuz in the Spanish-American War.[66] The story was planted and published by a Hearst newspaper, the *New York Journal*, and picked up by the *World*. The name of the distinguished "Austrian colonel" was an anagram of "we pilfer the news."[67] In a similarly notorious case occurring more than two decades later, the *New York City News Association* (*City News*) planted and circulated on its tickers a fake story about the arrest of one 'Nelson B. Steyne.' After the trap unmasked piracy on the part of *Illustrated News* (IN), which had rewritten and printed the story, the *City News* sought an injunction to prevent theft of its stories. The case ended with an apology and a dismissal of the suit.[68]

In *International News Service (INS) v. Associated Press (AP)*, decided in 1918, the U.S. Supreme Court criticized the behavior of INS—a Hearst news service that appropriated stories reported and written by the Associated Press (AP), rewrote them, and printed the stories without crediting the AP.[69] In dictum, authoritative language that is not binding, the court derided the practice of misappropriating and reusing news stories. "Besides the misappropriation," the court's decision proclaimed, "there are elements of imitation, of false pretense, in defendant's practices… The habitual failure to give credit to complainant for that which is taken is significant. Indeed, the entire system of appropriating complainant's news and transmitting it as a commercial product to defendant's clients and patrons amounts to a false representation to them and to their newspaper readers that the news transmitted is the result of defendant's own investigation in the field."[70] Although the journalism profession's attitude toward pilfering news stories had improved by the time of its promulgation, the American Society of Newspaper Editors' Canon of

Ethics, published in 1923, did not mention plagiarism or misappropriation of news stories.[71]

By 1951, however, *Times'* columnist and Pulitzer Prize winner Meyer Berger noted that while the early journalistic code was "a little crude and raw" there had been a change in journalism's attitude toward news piracy. "Property rights in news have been established through the years that have passed between," Berger observed, "and, instead of a reward for a stolen story, a Times [sic] man today would probably wind up with a reprimand, if not dismissal; the code had changed that much."[72]

Manuals of style from the early twentieth century had helped to codify guidelines for quoting, citing, and referencing the works of others over time. *The Chicago Manual of Style*, first published in 1906 under the title *Manual of Style Being a Compilation of the Typographical Rules in Force at the University of Chicago Press*, set forth rules for quoting material from outside sources: a quotation of three lines or less required quotation marks; longer quotations should be printed in a smaller font than the main text with no quotation marks; ellipses should be used to indicate omitted material.[73] The 1937 edition of the *Manual* (likely the edition in use when Laurence wrote *Dawn Over Zero*) outlined substantially similar requirements for quotations from outside sources,[74] and advised writers that "[a]ll extracts should correspond exactly with the original in wording, spelling, and punctuation."[75] Rules regarding footnotes[76] and bibliographies[77] were also included in the 1937 edition, making it clear that properly citing and referencing the works of authors was of the utmost importance in literary circles.

THE ONLY ONE WHO WAS THERE AND SAW IT ALL?

It's impossible to document with certainty Laurence's thought process as he incorporated portions of Siemes' material into *Dawn Over Zero* and *Men and Atoms*, but he left behind writings and an oral history that suggest his frame of mind during this period in his life. In an interview conducted in 1964, Laurence complained of his treatment at the *Times* following his work with the War Department. The *Times*, he groused, "had in many ways been really inconsiderate: (a) I expected to get a considerable increase in my salary—at least $100 a week increase because I was working for a small salary in those days... They [the *Times*], as I say, syndicated my stories and all that. I thought, you know, that I wouldn't have to ask, that it would be almost automatic. And to my great amazement, weeks, months passed, and I got no recognition whatever."[78] Laurence eventually received a small raise from the *Times*, but he thought it should have been larger and paid more promptly. The lack of a suitable raise was not Laurence's only grievance against the paper: "There was another thing. Immediately scores of publishers came after me to tell my story. I finally signed up with Alfred A. Knopf [also the publisher of the book version of Hersey's *Hiroshima*] with the fairly generous advance of $5000. But... the *Times* went and reprinted all these articles which told the entire story in a little booklet...and advertised it far and wide and sold it for 10 cents a copy. They sold 150,000 copies at least...by the time my book [*Dawn Over Zero*] came out, the *Times* had glutted the market with my original story... And so in that way, the *Times* ruined the sale of my book. It sold only a few thousand copies." "*Of course there were a lot of other fly-by-nights who rushed in to write books,*" Laurence added, "*...but I was the only one after all who was there and who saw it and all that* [emphasis provided]."[79] Despite Laurence having been there and seen it "and all that," by the time book arrived, it already faced stiff competition.

Dawn Over Zero was released by Knopf on August 22, 1946,[80] just days before John Hersey's "Hiroshima" was first published, in its entirety, in the August 31, 1946, issue of the *New Yorker*.[81] (It would appear in book format from Knopf on November 1, 1946.[82]) Correspondence between the *Times* and Knopf suggests that the publishing house was less than enthusiastic about promoting *Dawn Over Zero*, likely due to the imminent publication of *Hiroshima*. In an August 23, 1946, letter to *Times* editor Lester Markel, Mrs. Alfred Knopf, wrote: "I sent you down a copy of William L. Laurence's DAWN OVER ZERO... Is there anything that you can do for this book?" Mrs. Knopf asked. "After all he is an important Times [sic] man, and we are having hard sledding with it as you can imagine. I need all the help that we can get. Will you think it over [sic]."[83] On the same day, Blanche Knopf wrote the following to Arthur Hays Sulzberger: "I just sent the attached to Markel, and I am sending a copy of the book to you. I do not know whether we want to do anything or not, I hope there is something the Times [sic] can do."[84] The *Times* did in fact publish an editorial extolling Laurence's credentials as a science expert who had witnessed first-hand the events he described.[85] Yet when the *Times* published its Christmas book recommendations for 1946 on December 1, *Hiroshima* was on the list; *Dawn Over Zero* was not.[86]

Ironically, in 1948 Harry H. Moore edited a volume of reprinted essays, titled *Survival or Suicide: A Summons to Old and Young to Build a United, Peaceful World*.[87] The author of the first essay in the volume, which appeared under the chapter heading "A New Era Ushers in a New and Greater Crisis," was William L. Laurence. Sources of the essay's content were included in a footnote on the last page of the chapter. This footnote exemplifies the era's standard regarding permission to publish another author's work: "Reprinted from *Dawn Over Zero: the Story of the Atomic Bomb*, by William L. Laurence, by permission of Alfred A. Knopf, Inc. Copyright, 1946 by William L. Laurence; and from three articles in the *New York Times* December 1, 1946, by William L. Laurence, used with permission of author and publisher."[88]

WHY LAURENCE'S LAPSES MATTER

Critiquing Laurence's sourcing is more than just an academic exercise. Laurence's books, *Dawn Over Zero* and *Men and Atoms*, contained both original material and material derived from earlier works, including Laurence's *Times* stories and Father Siemes' reports. Siemes' eyewitness reports played a significant role in debate and discussion following the atomic bombing of Japan. The U.S. government was interested enough in Siemes' report to have Averill Liebow translate it and Siemes appeared in two post-war U.S. propaganda films in which he recounted sections of his report. The fact that Siemes appeared in these films suggests that the postwar occupiers of Japan wanted at least portions of Siemes' message to be disseminated. Not only did Laurence have access to and use Siemes' report or reports as a source, a version of Siemes' report appears to have played a role in the development and writing of Hersey's *Hiroshima*. The provenance of Siemes' eyewitness reports—how they were obtained and used—is thus historically important. Questions remain regarding the role the U.S. government played in connecting Laurence and Hersey with Siemes' reports.

As to Laurence, one must ask whether criticizing his imperfect acknowledgment of Siemes' work is presentist. Was Laurence a man of his time who adhered to then-current ethical values, or did he knowingly violate established norms regarding authorship, citing the works of others, and referencing? Teasing apart the writing and publishing ethics of Laurence's era is not easy, in part because of the overlap between copyright protection and plagiarism, and in part because journalistic ethics were evolving. As the works of a foreign author, Father Siemes' eyewitness statements likely did not fall within the purview of U.S. copyright law. Plagiarism, however, was clearly a concern of journalism during Laurence's era.

Laurence had no formal training in journalism. He was trained in the trenches. His actions as a translator and adaptor of foreign plays suggest that, from the beginning of his writing career, Laurence's authorial boundaries were questionable. He thought nothing of rewriting and adding his own flourishes to another author's work. Laurence's lack of a formal journalism education as well as the volatile ethical environment in which he found himself at the start of his career—a milieu in which newspapers were known to pilfer each other's stories and newspaper piracy cases were reported in the news—might have contributed to his lackadaisical attitude regarding the proper way to quote, cite, and reference another author's works. Laurence's seeming dismay at the manner in which his own byline was sometimes disregarded might have fostered an "everybody does it" attitude.

Based on statements Liebow made in his book, *Encounter with Disaster: A Medical Diary of Hiroshima, 1945*, it is likely that the Liebow translation was the first English translation of Siemes' account. By his own admission in *Dawn Over Zero*, Laurence clearly read the version of Siemes' report published in *Jesuit Missions*. Because of his relationship with the U.S. War Department, Laurence might also have had access to the Liebow translation. He also is likely to have read other versions of Siemes' report published before *Dawn Over Zero*. It is possible that the proliferation of several versions of the report in the months following the bombing, as well as the fact that the report was written by a foreign national and citizen of a recently conquered enemy government and was part of a trove of information that came into the possession of the War Department, could have led Laurence, as "historian" for the Manhattan Project, to believe he had free rein to use Siemes' materials however he wished.

On the other hand, around the time Laurence became a journalist, a revolution in journalism ethics had begun. Pilfering stories was frowned upon—not just from a commercial point of view, but also from an ethical perspective. A newspaper piracy case went all the way to the U.S. Supreme Court in 1918 and the *Times'* publisher, Adolphe Ochs, made it clear to Columbia journalism students that plagiarism was unacceptable. In his book, *The Story of the New York Times, 1851–1951*, Meyer Berger made similar representations about the ethical standard of Laurence's day.

In light of these facts, it is much more likely that Laurence, believing that he was the primary keeper of the story of the atomic bomb and wanting to capitalize on his unique experiences, wrote his two books with an eye toward their commercial potential, ignoring or minimizing the propriety of appropriating and rewriting another author's material. The elaborate transformation of Siemes' first-person eyewitness report into a third-person chronicle and the associated additions, deletions, and substitutions by Laurence suggest a level of engagement with the material that belies mere carelessness. All of Laurence's manipulations invite his readers, past and present, to think that he, Laurence, wrote portions of Siemes' frank and moving account of the devastation endured by the city of Hiroshima and its residents.

Notes

1. "William L. Laurence, Ex-Science Writer for the Times, Dies," *New York Times*, March 19, 1977, 1, 7.

2. "Pulitzer Prizes Awarded, 'State of the Union' Pulitzer Winner," *New York Times*, May 7, 1946, 1, 14.

3. Amy Goodman and David Goodman, "Hiroshima Cover-up: How the War Department's *Times* Man Won a Pulitzer," *Common Dreams*, Aug. 10, 2004; Amy Goodman and David Goodman, *The Exception to the Rulers: Exposing Oily Politicians, War Profiteers, and the Media That Love Them* (New York: Hyperion, 2004); Beverly Deepe Keever, *News Zero: The New York Times and the Bomb* (Monroe, ME: Common Courage Press, 2004); Robert J. Lifton and Greg Mitchell, *Hiroshima in America: A Half Century of Denial* (New York: Harper Perennial, 1995).

4. William L. Laurence, *The Story of the Atomic Bomb* (Rockville, MD: Wildside Press, 2009); William L. Laurence, "U.S. Atom Bomb Site Belies Tokyo Tales," *New York Times*, September 12, 1945, 1, 4; William L. Laurence, "No Radioactivity In Hiroshima Ruin," *New York Times*, September 13, 1945, 4; Wilfred Burchett, "The Atomic Plague: I Write This as a Warning to the World," *The Daily Express*, September 5, 1945,

1; George Burchett and Nick Shimmin, eds. *Rebel Journalism: The Writings of Wilfred Burchett* (Cambridge University Press, 2007).

5. William L. Laurence, "Vast Power Source in Atomic Energy Opened by Science," *New York Times*, May 5, 1940, 1, 51; William L. Laurence, "The Atom Gives Up," *Saturday Evening Post*, September 7, 1940,12–13, 60–63; William L. Laurence, *Dawn Over Zero: The Story of the Atomic Bomb* (New York: Knopf, 1946); William L. Laurence, *Men and Atoms: The Discovery, the Uses, and the Future of Atomic Energy* (New York: Simon and Schuster, 1959); David W. Moore, "Majority Supports Use of Atomic Bomb on Japan in WWII," *Gallup News Service*, August 5, 2005, available at http://news.gallup.com/poll/17677/majority-supports-use-atomic-bomb-japan-wwii.aspx; Bruce Stokes, "70 Years After Hiroshima, Opinions Have Shifted on Use of Atomic Bomb," *Pew Research Center*, August 4, 2015, available at http://www.pewresearch.org/fact-tank/2015/08/04/70-years-after-hiroshima-opinions-have-shifted-on-use-of-atomic-bomb/.

6. Laurence, *Dawn Over Zero*, 245–250; Laurence, *Men and Atoms*, 163–178; John A. Siemes, S.J., "The Atomic Age: Hiroshima Eye-Witness," *The Saturday Review of Literature*, May 11, 1946, 24–25, 40–44. Also see the Mendeley dataset which sets forth tables containing relevant Laurence citations, citations for versions of Siemes' eyewitness report on Hiroshima, page references for Laurence's plagiarism as well as figures illustrating specific textual examples of Laurence's plagiarism. https://data.mendeley.com/datasets/y4ffj425jw/draft?a=56430aa0-4c24-4c24-8e3b-124503dc38cc.

7. Laurence, *Dawn Over Zero*, 245–249; Laurence, *Men and Atoms*, 163–178; Siemes, "The Atomic Age: Hiroshima Eye-Witness," 24–25, 40–44. To make a comparison of the texts, see also the tables and figures contained in the Mendeley dataset for examples of this transformation: https://data.mendeley.com/datasets/y4ffj425jw/draft?a=56430aa0-4c24-4c24-8e3b-124503dc38cc.

8. Siemes, "The Atomic Age: Hiroshima Eye-Witness," 24.

9. Siemes, "Atomic Bomb on Hiroshima, eyewitness account of P. [sic] Siemes," trans. Averill Liebow, in Averill A. Liebow Collection, MS Coll 28, Box 3, Folder 36, Harvey Cushing/John Hay Whitney Medical Library, Yale University; Siemes, "The Atomic Age: Hiroshima Eye-Witness," 24–25, 40–44.

10. *The Atom Strikes!* U.S. Army Pictorial Services, U.S. Army Signal Corps, U.S. War Department (1945), available at https://www.youtube.com/watch?v=C9VrMfAUMX4; *A Tale of Two Cities*, Army Navy Screen Magazine, War Department, available at https://www.youtube.com/watch?v=H2lIad6WLlI.

11. See Table 3 in the "Borrowed Chronicles" dataset at https://data.mendeley.com/datasets/y4ffj425jw/draft?a=56430aa0-4c24-4c24-8e3b-124503dc38cc.

12. Averill A. Liebow, *Encounter with Disaster: A Medical Diary of Hiroshima* (New York: W.W. Norton, 1970), 82–83.

13. Ibid., 17–20.

14. Ibid., 82.

15. Siemes, "Atomic Bomb on Hiroshima, eyewitness account of P. [sic] Siemes," trans. Averill Liebow, in Averill A. Liebow Collection, MS Coll 28, Box 3, Folder 36, Harvey Cushing/John Hay Whitney Medical Library, Yale University; Liebow, *Encounter with Disaster: A Medical Diary of Hiroshima*, 82–83.

16. Siemes, "The Atomic Age: Hiroshima: Eye-Witness," 24–25, 40–44. Siemes, "Atomic Bomb on Hiroshima, eyewitness account of P. [sic] Siemes," trans. Averill Liebow, in Averill A. Liebow Collection, MS Coll 28, Box 3, Folder 36, Harvey Cushing/John Hay Whitney Medical Library, Yale University; Liebow, *Encounter with Disaster: A Medical Diary of Hiroshima*, 82–83.

17. Laurence, *Dawn Over Zero*, vii–viii, 245–249. See also the tables, figures, and supplemental material contained in the Borrowed Chronicles dataset at https://data.mendeley.com/datasets/y4ffj425jw/draft?a=56430aa0-4c24-4c24-8e3b-124503dc38cc.

18. Laurence, *Men and Atoms*, 163–178. See also the tables, figures and supplemental material contained in the Borrowed Chronicles dataset at https://data.mendeley.com/datasets/y4ffj425jw/draft?a=56430aa0-4c24-4c24-8e3b-124503dc38cc.

19. William L. Laurence, "Atomic Bombing of Nagasaki told by Flight Member," *New York Times*, September 9, 1945, 1; Laurence, *Dawn Over Zero*, 228–238; Laurence, *Men and Atoms*, 154–160.

20. J. Siemes, "Report From Hiroshima," *Jesuit Missions* 20, no. 2 (1946): 30–32; Siemes, "The Atomic Age; Hiroshima Eyewitness," *The Saturday Review of Literature*, May 11, 1946, 24–25, 40–44; Laurence, *Dawn Over Zero*, vii–viii, 245–249. See also representative excerpts from *Dawn Over Zero*, compared to the version of Siemes' report published in *Jesuit Missions* or *The Saturday Review of Literature*, which can be found in figures 1, 2, and 3 of the Mendeley dataset associated with this essay. These comparisons represent only a portion of the material Laurence appropriated from Siemes. Additional copied and manipulated passages appear throughout chapter eighteen of *Dawn Over Zero*.

21. Laurence, *Dawn Over Zero*, vii–viii.

22. Laurence, *Men and Atoms*, 163–178. Laurence interspersed Siemes' material, quoted nearly word-for-word, with his own text about the daily observation plane that flew over Hiroshima just prior to the bombing. See Figure 4 in the Mendeley dataset associated with this essay. The dataset is available at https://data.mendeley.com/datasets/y4ffj425jw/draft.

23. Ibid.

24. Laurence, *Dawn Over Zero*, vii–viii.

25. "William Laurence, Ex-Science Writer for the Times, Dies," *New York Times*, March 19, 1977, 1, 7.

26. William L. Laurence, tape recorded interview by L.M. Starr, March 27, 1956, transcript (Oral History Research Office, Columbia University), 1–15.

27. Ibid., 13.

28. "William L. Laurence, Ex-Science Writer for the *Times*, Dies," *New York Times*, March 19, 1977, 1, 7.

29. Ibid.

30. Maxim Gorki, *At the Bottom*, trans. William L. Laurence (New York: S. French, 1930); Maksim Gorky, *The Lower Depths, A drama in four acts* trans. Jenny Covan (1922, Neuilly sur Seine: Ulan Press, 2012); "The Theatre: Revivals," *Time*, January 20, 1930, 34; "At the Bottom," Internet Broadway (IBDB), available at https://www.ibdb.com/broadway-production/at-the-bottom-11027.

31. Arthur Pollock, "Maxim Gorky's 'At The Bottom' Has Broadway Premiere–Other Theater News," *The Brooklyn Daily Eagle*, January 10, 1930, 23.

32. "Devil in the Mind," Internet Broadway Database (IBDB), Available at https://www.ibdb.com/broadway-production/devil-in-the-mind-11360; "Theatre: New Play in Manhattan," *Time*, May 11, 1931, 72.

33. Justin Spafford and Lucien Esty, *Ask Me Another! The Question Book* (New York: The Viking Press, 1927); William L. Laurence, tape recorded interview by Scott Bruns, March 6, 1964, transcript (Oral History Research Office, Columbia University), 158–160.

34. Ibid. 160.

35. Ibid. 173.

36. Ibid. 176–180.

37. Ibid. 180.

38. Ibid. 179–182.

39. Ibid. 180–190; Laurence, *Dawn Over Zero*; Laurence, *Men and Atoms*.

40. Ibid.

41. "The Pulitzer Prizes," *New York Times*, May 5, 1937, 24; *The Pulitzer Prizes*, available at http://www.pulitzer.org/winners/john-j-oneill-william-l-laurence-howard-w-blakeslee-gobind-behari-lal-and-david-dietz.

42. "War Department Called *Times* Reporter to Explain Bomb's Intricacies to Public," *New York Times*, August 7, 1945, 5.

43. Ibid.

44. Op. cit.

45. Laurence, *Story of the Atomic Bomb*, 10–17; Laurence, *Dawn Over Zero*, 179–195; Laurence, *Men and Atoms*, 115–120.

46. Ralph Engelman, *Friendly Vision: Fred Friendly and the Rise and Fall of Television Journalism* (New York: Columbia University Press, 2009), 64–69; *The Quick and the Dead*, four-part series on atomic energy, producer Fred Friendly, host Bob Hope, narrator Robert Trout, supervisor William F. Brock (NBC, July 7–27, 1950).

47. William L. Laurence, "Atomic Bombing of Nagasaki Told by Flight Member," *New York Times*, September 9, 1945, 1. General Leslie M. Groves, *Now It Can Be Told: The Story of the Manhattan Project*, (New York: Harper & Row, 1962), 327; Laurence, *Story of the Atomic Bomb*; Laurence, *Men and Atoms*, 111–114.

48. William L. Laurence, "Eyewitness Account Atomic Bomb Mission Over Nagasaki," War Department Press Release, Sept. 9, 1945, available at http://www.atomicarchive. com/Docs/Hiroshima/Nagasaki.shtml.

49. Laurence, tape-recorded interview, May 9, 1964, transcript, 394. In his 1964 reminiscences, Laurence bemoaned the fact that some of his stories about the atomic bomb were rewritten and released under other journalists' bylines. "I had the curious experience of watching my own story in my own paper with a byline of somebody else. Well, I can say that he didn't exactly use [my] story word for word, but most of the material was simply a patched up job on the articles I had prepared."

50. Amy Goodman and David Goodman, "The Hiroshima Cover-up: How The Military Suppressed Early Reporting on the Atomic Devastation in Japan–with Help from *The New York Times*," *Mother Jones*, Aug. 5, 2005, http://www.motherjones.com/ politics/2005/08/hiroshima-cover/; Amy Goodman and David Goodman, *The Exception to the Rulers: Exposing Oily Politicians, War Profiteers, and the Media That Love Them* (New York: Hyperion, 2004); Beverly Deepe Keever, *News Zero: The New York Times and the Bomb* (Monroe, ME: Common Courage Press, 2004); Robert J. Lifton and Greg Mitchell, *Hiroshima in America: A Half Century of Denial* (New York: Harper Perennial, 1995).

51. Lisa K. Fazio, Nadia Brashier, B. Keith Payne, Elizabeth J. Marsh, "Knowledge Does Not Protect Against Illusory Truth," *Journal of Experimental Psychology* 144, no. 5 (2015): 993–1002.

52. Lisa Ede, "The concept of authorship: An historical perspective" (Paper Presented at the Annual Meeting of the National Council of Teachers of English, ERIC, 1985, 5), available at http://files.eric.ed.gov/fulltext/ED266481.pdf; Elizabeth L. Eisenstein, *The Printing Press as an Agent of Change* (Cambridge, UK: Cambridge University Press, 1979), 1:229–230.

53. U.S. Const. art. I, §8, cl. 2.

54. 28 U.S. Code § 2338 et. Seq. Melville Nimmer et al., eds. Eighth Ed. *Cases and Materials on Copyright and Other Aspects of Entertainment Litigation Including Unfair Competition, Defamation, Privacy* (San Francisco: Matthew Bender, 2012).

55. Copyright Act of 1790, 1 Stat. 124, available at https://www.copyright.gov/ history/1790act.pdf; Association of Research Libraries, *Copyright Timeline: A History of Copyright in the United States*, available at http://www.arl.org/focus-areas/copyright-ip/2486-copyright-timeline#.WcwHaNOGPBJ.

56. Orin G. Hatch, "Better Late Than Never: Implementation of the 1886 Berne Convention," *Cornell International Law Journal* 22, no. 2 (1989): 171–195.

57. 17 U.S. Code §105-Subject matter of copyright: United States Government works; "About U.S. Government Works," USA.gov, available at https://www.usa.gov/ government-works.

58. Roger Billings, "Plagiarism in Academia and Beyond: What is the Role of the Courts?" *University of San Francisco Law Review* 38, no. 3 (Spring 2004): 391–430.

59. Rebecca Moore Howard, "Plagiarisms, Authorships, and the Academic Death Penalty," *College English* 57, no. 7 (1995): 788–806; Rebecca Moore Howard, "A Plagiarism Pentimento," *Journal of Teaching Writing* 11.3 (Summer 1993): 233–246; Michele Eodice, "Plagiarism, Pedagogy, and Controversy: A Conversation with Rebecca Moore Howard," *Issues in Writing* 13, no. 1 (2002): 6–26.

60. Billings, 392, citing Robert Gorman and Jane C. Ginsburg, *Copyright: Cases and Materials*, 6th ed. (New York, 2002).

61. Richard A. Posner, *The Little Book of Plagiarism* (New York: Pantheon, 2007), 12.

62. Ibid., 12–13.

63. Ibid. 16–17.

64. John Higham, "Habits of the Cloth and Standards of the Academy," *The Journal of American History* 78, no. 1 (1991), 108.

65. Jack Shafer, "Don't get all huffy about the *Huffington Post*: The sort of borrowing it does is in the American journalistic tradition," *Slate*, April 15, 2009; "We apologize for what we said of one Nels B. Steyne," *Editor and Publisher*, May 15, 1920, 31; "Injunction suit claims 'piracy:' New York City News Association sues *Illustrated News* on basis of alleged 'fake' story," *The Fourth Estate*, May 15, 1920, 4; Meyer Berger, *The Story of the New York Times: 1851–1951* (New York: Simon and Schuster, 1951), 321.

66. "Topic of the Times," *New York Times*, June 10, 1898, 6.

67. Ibid.

68. "We apologize for what we said of one Nels B. Steyne," *Editor and Publisher*, May 15, 1920, 31; "Injunction suit claims 'piracy:' *New York City News Association* sues *Illustrated News* on basis of alleged 'fake' story," *The Fourth Estate*, May 15, 1920, 4.

69. *International News Service* v. *Associated Press*, 248 U.S. 215 (1918).

70. Ibid. 242.

71. American Society of Newspaper Editors' Code of Ethics or Canons of Journalism (1923), *Illinois Institute of Technology, Ethics Code Collection*, available at http://ethics.iit.edu/ ecodes/node/4457.

72. Meyer Berger, *The Story of the New York Times, 1851-1951* (New York, NY: Simon and Schuster, 1951), 21.

73. *Manual of Style Being a Compilation of the Typographical Rules in Force at the University of Chicago Press* (Chicago: University of Chicago Press, 1906), 25–28.

74. *A Manual of Style: Tenth Revised Edition with Specimens of Type.* (Chicago: University of Chicago Press, 1937).

75. Ibid. 52.

76. Ibid. 123–130.

77. Ibid. 131–132.

78. Laurence, tape-recorded interview, transcript (May 9, 1964), 403–404; Groves, *Now It Can Be Told*, 326. General Leslie Groves wrote that both the War Department and the *Times* paid Laurence while he was embedded with the War Department.

79. Laurence, tape-recorded interview, transcript (May 9, 1964), 404–405.

80. "Books Published Today," *New York Times*, August 22, 1946, 38.

81. Hersey, John, "Hiroshima," *New Yorker*, August 31, 1946, 15.

82. "Hersey's Hiroshima Report Now Published in Book Form," *Honolulu Star-Advertiser*, November 3, 1946, 55.

83. August 23, 1946, letter from Mrs. Alfred Knopf to *Times* editor, Lester Markel, *New York Times* Company records. Arthur Hays Sulzberger papers. Manuscripts and Archives Division. The New York Public Library. Astor, Lenox, and Tilden Foundations.

84. August 23, 1946, letter from Blanche Knopf to Arthur Hays Sulzberger, *New York Times* Company records. Arthur Hays Sulzberger papers. Manuscripts and Archives Division. The New York Public Library. Astor, Lenox, and Tilden Foundations.

85. "Dawn Over Zero," *New York Times*, September 10, 1946, 6.

86. John Chamberlain, "Ten Christmas Lists of '10 Best,'" *New York Times*, December 1, 1946, 94.

87. William L. Laurence, "A New Era Ushers in a New and Greater Crisis," in *Survival or Suicide: A Summons to Old and Young to Build a United, Peaceful World*, ed. Harry Hascall Moore (Freeport, NY: Books for Libraries Press, 1971), 3–7.

88. Ibid.

NECESSITY IS THE MOTHER OF INVENTION

CHAPTER THREE

Casting Shadows, Capturing Images
The History and Legacy of Implosion Physics at Los Alamos

Ellen D. McGehee

INTRODUCTION

At Los Alamos, scientists and engineers tackled one of the Manhattan Project's greatest scientific hurdles: the unsuitability of the plutonium gun device and the need to develop an alternate weapon design. Buildings and artifacts associated with wartime implosion diagnostic work remain at present-day Los Alamos National Laboratory, and several of the implosion test sites are included in the 2014 legislation that authorized the newly established Manhattan Project National Historical Park at three sites across the country: Los Alamos, New Mexico; Oak Ridge, Tennessee; and Hanford, Washington. This essay describes the history and legacy of implosion physics at Los Alamos, beginning with an exploration of the Manhattan Project-era experimental approaches that eventually led to the design of the Fat Man weapon used against Japan. The present-day memory and meaning of Los Alamos's wartime places and spaces of the bomb is examined through the lens of scientific, social, and environmental themes recently developed by the National Park Service for the new park.

PART I: HISTORY
Project Y: The Secret Laboratory at Los Alamos

The Manhattan Project was an unprecedented, top-secret government program carried out during World War II with the primary goal of constructing a nuclear bomb before Nazi Germany did. The wartime

51

program resulted in science and technology that transformed the role of the United States in the world and ushered in the atomic age. As part of the Manhattan Project, a secret laboratory was established at Los Alamos, on the high mesas of the Pajarito Plateau in northern New Mexico. Code-named Project Y, the Los Alamos laboratory was made up of a diverse group of workers, including scientists, technicians, and members of the military, whose singular focus was to build an atomic weapon to end the war. In the early months of Project Y, designing a gun-assembled bomb—essentially the shooting of one subcritical mass at another—was given top priority. Plans involved the development of two gun-assembled weapons, one (code-named Thin Man) that could use the plutonium coming from Hanford, and the other (code-named Little Boy) using the uranium coming from Oak Ridge.

Top scientists at Los Alamos initially relegated the development of an "implosion" assembled weapon (code-named Fat Man) to the wings, and rudimentary field tests of the implosion concept were conducted by physicist Seth Neddermeyer from the California Institute of Technology at a remote location south of the Los Alamos townsite.[1] Neddermeyer's early tests with cylinders revealed that achieving a symmetrical implosion (or inward explosion) would present a challenging task; he and other Project Y researchers rapidly understood that a key problem to resolve was how to see inside an explosive event. The implosion weapon's high explosives lenses would need to be simultaneously detonated to compress a subcritical mass of plutonium, causing it to go supercritical. Ultimately, as part of their implosion research, the scientists needed to understand the velocity, symmetry, and compression of prototype high-explosives assemblies.[2]

A more permanent installation for early implosion studies and high-explosives research was established on the Los Alamos site in October 1943, at Anchor Ranch East, later known as Technical Area (TA) 9. There, small-scale tests were conducted by Neddermeyer and other scientists, including Kenneth Greisen, leader of the flash-X-ray photography effort, and Joseph Hoffman and Walter Koski, pioneers in Los Alamos's high-speed camera research.[3] Early experiments relied on explosive flash and X-ray methods to produce diagnostic images on film. High-speed cameras were also essential tools used for these first tests, which involved the detonation of small spheres of high explosives in the hopes of gathering data on the velocity and symmetry of implosions.[4]

The Crisis of 1944

In 1944, a group of scientists working at remote Pajarito Site under Emilio Segrè determined that plutonium could not be used in the Thin Man design because the plutonium contained an isotope (plutonium-240) that released neutrons. This high-neutron background would cause the nuclear chain reaction to start prematurely if an assembly method as slow as the gun device was used.[5] The realization that plutonium could not be used in the gun device led to the abandonment of the Thin Man weapon, resulting in what is now known as "the crisis of '44." In response, J. Robert Oppenheimer reorganized Project Y on August 14, 1944, with one goal in mind: to develop an implosion or Fat Man-type weapon to make use of Hanford's plutonium.[6]

Hundreds more workers came to Los Alamos to support detonator development, high-explosives production, and explosives tests. George Kistiakowski, a Ukrainian-born high-explosives expert, was tasked to lead the effort to design the special high-explosives lenses that were an integral component of the Fat Man weapon.[7] Remarkably, at least seven diagnostic testing methods were developed to study the inner workings of implosion, and twelve new technical areas (including firing sites and remote laboratories) were established in response to the crisis of 1944.[8] Most of these wartime implosion test sites were located south of town, and many of these sites are still located within the active mission areas of present-day Los Alamos National Laboratory (Figure 3.1).

The Implosion Problem

Several factors contributed to the problems associated with the development of a completely new method of weapon assembly. The initial implosion studies conducted at Anchor Ranch East were only just beginning to provide consistent data at the time of Oppenheimer's reorganization, and new methods of studying implosion were not yet fully developed or were only possible in theory. Moreover, high-explosives and detonator systems, key to producing a symmetrical inward explosion, were still in the preliminary design stages at this time.[9]

A diversity of methods had to be used at Los Alamos to overcome the implosion problem, from the dedicated use of outlying technical areas for implosion testing to the use of overlapping diagnostic studies

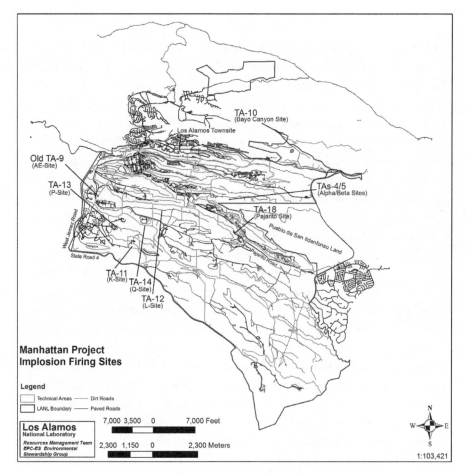

3.1. Map of Manhattan Project implosion firing sites. *Courtesy of Los Alamos National Laboratory.*

exemplified by the simultaneous development of the terminal observation, electric pin, magnetic, X-ray, betatron, and RaLa methods.[10] Scientists and engineers were rapidly developing the specialized detection and recording equipment—ionization chambers, cathode ray oscilloscopes, and armored high-speed cameras—needed to document the static and dynamic conditions that were part of the high-explosives tests associated with the new field of implosion physics.[11]

More difficult to perfect than the uranium gun-assembled design, the plutonium implosion design was extremely complex, and its success was fraught with uncertainty because the implosion weapon called for new technology and materials that had not yet been invented. Italian scientist Bruno Rossi, writing in his memoir about wartime Los Alamos, recalled the challenges posed by the implosion concept: "[T]he development of the implosion method involved many more serious problems than the development of the gun method—problems whose solutions were likely to require more time than was available."[12] He also reflected that "implosion was a 'new' phenomenon, one which had never been studied either experimentally or theoretically."[13] Initial implosion work at the new firing sites focused on small-scale tests using more or less standard diagnostic techniques. Small-scale implosion "assemblies" consisted of high-explosives charges using non-fissile metallic pits as stand-ins and could be produced in mass quantities for the round-the-clock tests required to achieve the rapid design progress demanded by wartime urgency. Technical areas at Los Alamos were typically given letter designations during the war instead of the site numbers in use today. Many of the new implosion sites were assigned the letter that corresponded with the last name of the group leader in charge of site operations, hence P-Site was named for Lyman Parratt, L-Site was named for Henry Linschitz, K-Site for Donald Kerst, and so on.

Electric Pin

Electric pin tests were carried out under the direction of group leader Darol Froman at Alpha and Beta Sites, located north of Pajarito Road at present-day TAs 4 and 5. The tests, conducted in a firing pit, documented the symmetry and velocity of an implosion. In this method, first developed in August of 1944, changes in electrical impulses were detected as pins were struck by the inward movement of metal pieces used in the shot assemblies, and shorts to electrical signals were recorded for each contact. Scientists used oscilloscopes to capture the electronic record; a nearby battery building also provided the power needed for the electric pin tests.[14]

Terminal Observation

Terminal observation was another early method that Los Alamos scientists employed to research the concept of implosion. Initially developed in October of 1944 at a firing site known as L-Site, the terminal observation method involved the detonation of small-scale, high-explosives assemblies inside a large, steel-lined pit with a metal cover. The remote test site also included a hutment building and several small explosives magazines. After each firing test, scientists from Los Alamos group X-1B examined the resultant shot debris and also analyzed the damage to specially placed steel slabs to diagnose the success of implosion tests. Henry Linschitz was the group leader in charge of L-Site; key personnel also included scientist Lilli Hornig, who analyzed shot data and authored some of the group's progress reports.[15]

Magnetic Method

The magnetic method, in use at Pajarito Site beginning in early 1945, was based on the principle that movement of metal in a magnetic field disrupts the field. Following this principle, the inward motion of a metal core imploded during a high-explosives test was detected when it created a current in a nearby magnetic pickup coil (Figure 3.2). Oscilloscopes and high-speed cameras were used at the site to record test data. Special concrete control-room buildings called Battleship Bunkers housed

3.2. Magnetic method experimental setup. *Courtesy of Los Alamos National Laboratory.*

personnel and test equipment. Nearby battery buildings and generators provided the power for the shots. The magnetic method experiments, conducted under the leadership of Edwin McMillan, provided information related to velocity and symmetry and, most importantly, could be used for full-scale assemblies.[16]

X-ray Methods

Several of the firing sites relied on X-ray sources that could generate short pulses of energy during an implosion test. These X-ray machines were paired with various detection methods that could document the baseline or static conditions before a test and the dynamic conditions occurring during a test.[17]

P-Site (named after Lyman Parratt) was in use from the fall of 1944 until March of 1945 for flash-X-ray and counter experiments. At P-Site, X-rays aimed at an implosion shot were captured as a shadow image by small Geiger counters arranged in an array on a detector post. Initially promising, this method was abandoned in the spring of 1945 in favor of more productive methods, such as the RaLa and betatron methods.[18] Site workers included group leader Lyman Parratt, a recognized expert in high-precision X-ray measurements, and British physicist James Tuck (Figure 3.3). The primary work at P-Site relied on an X-ray source—in this case, Westinghouse X-ray tubes—in conjunction with variously scaled shots that were situated on a firing pad near the nose of a concrete bunker building housing the X-ray equipment. The cross-shaped portion of a detector post was oriented in the X-ray shadow of the blast and was designed to hold individual Geiger counters aligned inside the arms of the cross. As proposed, a simple detect or non-detect signal by the individual counters would generate the data. Circuitry, electronics, and a power-source for the X-ray equipment were key supporting technologies.[19]

K-Site was constructed in October of 1944 for implosion experiments using a betatron machine. Donald Kerst from the University of Illinois was the inventor of the betatron machine and led the research efforts at K-Site. Key facilities associated with the betatron method included two nose-to-nose concrete bunker buildings, one housing Donald Kerst's betatron (a gamma ray emitter) and the other containing a cloud chamber and high-speed camera. For each experimental shot, a high-explosives assembly

3.3. P-Site workers, with James Tuck at far right. *Courtesy of Los Alamos National Laboratory.*

was detonated on a shot pad located between the two buildings, whose front sections were armored for protection. Betatron radiation passing through the implosion created a shadow image, which could be captured using the cloud chamber and camera located in the bunker opposite the betatron. The betatron machine's radiation was more penetrating than the flash X-rays generated at P-Site, and the betatron and cloud chamber method eventually became the preferred X-ray technique in use during the Manhattan Project. Although some aspects of this method had to be developed at Los Alamos, Seth Neddermeyer, who pioneered the implosion concept in 1943, had been conducting research using cloud chambers before the war while at Caltech with Nobel-prize winning physicist Carl Anderson.[20]

RaLa Method

Small-scale implosion research began at Los Alamos in 1943 with Seth Neddermeyer's field experiments; however, by the winter of the same year, a dedicated group of Los Alamos scientists was actively engaged in making mathematical calculations that would aid in the understanding of implosion physics.[21] While most of the other wartime diagnostic methods were useful in testing the theorists' implosion models, one method stands out for its ability to capture the total continuum of a test and not just a series of data snapshots: this was the RaLa (radiolanthanum) method.[22] As envisioned by Los Alamos scientist Robert Serber, the experiment would make use of a RaLa source (a powerful gamma emitter), which would be placed at the heart of an implosion assembly. Fast ionization chambers would be located near the test setup to detect the radiation being released from the experiment (Figure 3.4). Once detonated, the lenses of high explosives would compress the core of the assembly, increasing its thickness, and reducing the amount of radiation being detected. Signals from the radiation detectors would be transmitted to oscillographs making continuous sweeps from the moment of detonation; these data traces would then be recorded using high speed cameras.[23]

3.4. RaLa shot at Bayo Canyon Site. Ionization chambers at left and right. *Courtesy of Los Alamos National Laboratory.*

As part of his Los Alamos duties, scientist Bruno Rossi had worked on developing specialized detectors for some of the early weapons-design experiments.[24] Rossi had partnered with physicist Hans Staub in founding the Detector Group (P-6), which was part of Robert Bacher's Physics Division.[25] Rossi noted that "the most important result of our work on instrumentation for the general use of the laboratory was the development of the fast ionization chamber."[26] This instrumentation would become a critically important tool to detect small changes in radiation during experimental tests and was able to follow these changes as they were rapidly occurring during explosive events.[27] Because the Detector Group was involved in developing fast ionization chambers, Rossi and Staub were logical choices to lead the RaLa research—first working with physicist Luis Alvarez, who had come to Los Alamos from the University of California at Berkeley, and later assuming all responsibilities for the program.[28]

RaLa implosion tests were conducted in remote Bayo Canyon, north of Los Alamos's main laboratory and residential area. The longest lived of the wartime implosion experiments, RaLa shots were conducted from October of 1944 to the early 1960s. Radiobarium from the Oak Ridge pile was the source of radioactivity for the test shots, and, amazingly, the radiolanthanum was separated from the radiobarium in a primitive wood-frame laboratory building at the Bayo Canyon Site. Women's Army Corps (WAC) chemist Norma Gross was a member of the radiochemistry team. In preparation for the tests, chemists would place the RaLa source into a lead container, effectively shielding the source during transport to the firing area. Rossi, in his memoir, recalled how site worker Benjamin Diven had "set up a system of pulleys and strings which allowed him to perform all the necessary operations without coming closer than twelve meters to the radioactive source."[29] After experiments were discontinued, the remaining site facilities were removed in 1963.

PART II: LEGACY

Science: Implosion Physics—The Cold War and Beyond

There are many legacies of the Manhattan Project: scientific, technical, military, geopolitical, social, cultural, human health, and environmental.[30] Los Alamos's role in the Manhattan Project did not conclude with the end of World War II. The refinement and testing of weapon designs continued at Los Alamos, and postwar weapons tests began in 1946 in the Pacific. After the war, General Groves tasked Los Alamos with the production of the country's first atomic stockpile—weapons components would be produced at Los Alamos, but weapon assembly would be carried out at nearby Sandia Base in Albuquerque.[31] The demand for Cold War weapon designs resulted in the growth of the U.S. nuclear testing program. Nonnuclear tests of key weapons components and the eventual need to certify the nation's stockpile in the absence of nuclear testing drove home the continued importance of implosion physics—a new field of high-energy research first developed at the small wartime Los Alamos firing sites.[32] Most implosion experiments, such as the terminal observation, X-ray and counter, and magnetic methods, were discontinued by the end of the war, and the RaLa implosion experiments were terminated in the early 1960s. However, flash X-ray machines and high-speed cameras, technology born of the Manhattan Project, continued as fundamental diagnostic tools used to support implosion research and stockpile management throughout the Cold War years.

PHERMEX

A key post-war facility, PHERMEX, short for Pulsed High-Energy Radiographic Machine Emitting X-rays, was constructed at Los Alamos in 1962 to conduct nonnuclear implosion tests. PHERMEX became even more important to the Cold War management of the nuclear stockpile after the 1963 Limited Test Ban, which ended atmospheric testing. Then laboratory director Norris Bradbury, speaking at a 1963 news conference, summed up some of the technological solutions that could be used by those managing the nation's stockpile in the absence of aboveground nuclear testing, emphasizing nonnuclear tests of weapon components that would become even more significant to Los Alamos in 1992 with the cessation of all U.S. nuclear tests. In 1963, Bradbury acknowledged that "these are very involved technical systems, atomic weapons; they have new materials, strange materials, exotic materials, requirements of very great precision of character and behavior," adding that "one can test it in some cases piecewise. One can find out things about the system without necessarily resorting to a complete, full-blown...nuclear explosion."[33]

The PHERMEX machine, like the Manhattan Project-era X-ray machines, could produce a series of X-ray images taken during implosion tests.[34] Relying on the same flash radiographic technology in use at Anchor Ranch East, P-Site, and K-Site during the war, the PHERMEX machine was an important facility at the post-war laboratory for over

30 years (Figure 3.5). During this time, Los Alamos scientists used the machine to study the hydrodynamic behavior of large explosively driven assemblies. PHERMEX was much more powerful than its Manhattan Project predecessors, and its 27-MeV linear electron accelerator could produce precise and penetrating pulses of radiation.[35]

3.5. The PHERMEX machine. *Courtesy of Los Alamos National Laboratory.*

DARHT

At Los Alamos, science-based stockpile management became a necessity after the end of U.S. nuclear testing in 1992. Completed in 1999, the DARHT (Dual-Axis Radiographic Hydrodynamic Test) facility houses a powerful X-ray machine that analyzes mockups of nuclear weapons by generating a series of three-dimensional images during high-speed implosion tests (Figure 3.6). In order to capture the radiographs, the DARHT facility makes use of two electron accelerators placed at right angles to each other to create a powerful X-ray burst.[36] Present-day stockpile stewardship research, exemplified by the work conducted at Los Alamos's DARHT facility, owes its scientific origins to Project Y of the Manhattan Project.

Memory and Place: The Archaeology of the Manhattan Project

Only a handful of buildings and structures associated with the wartime laboratory at Los Alamos remain. Manhattan Project landscapes have been dramatically transformed into modern research areas, and many of the implosion test areas that supported the development of the Fat Man weapon are little more than archaeological sites. However, these lost places of World War II still retain historical significance, and the interpretation of the scientific landscapes of the bomb can serve to stimulate meaningful dialogues about the past, its memory, and meaning. At present-day

3.6. The DARHT facility. *Courtesy of Los Alamos National Laboratory.*

Los Alamos National Laboratory, archaeological research at some of the abandoned wartime sites provides a greater understanding of the specific technologies born of the war and, through the study of the material culture of implosion physics, reveals important scientific and engineering information not found in Project Y's document archives.

Pajarito Site and the Creutz Test

The Creutz implosion test was the final systems check of the Trinity test device, code-named "the Gadget." Located at Pajarito Site in remote Pajarito Canyon, the Creutz test was conducted on July 14, 1945, just two days before Trinity. Named for Edward Creutz, then leader of Los Alamos's Magnetic Method Group, the test made use of the magnetic method to document the symmetry and velocity of the Creutz implosion because it was the only method that could be used for full-scale assemblies. The Creutz test used high-explosives lenses identical to those made for the Trinity test. However, a non-fissile metal pit was substituted, and the surrounding case was made of plastic instead of metal.[37]

The results of that dress rehearsal, without the active fissile material, were initially disappointing and suggested that the test of the Trinity Gadget would fail. Hans Bethe's analysis of the results, though, indicated that the Creutz test worked, much to the relief of those preparing the test of the

3.7. Aftermath of the Creutz test, July 14, 1945. *Courtesy of Los Alamos National Laboratory.*

implosion device already at the tower at Trinity Site.[38] The firing site at Pajarito Canyon was abandoned after the war, and the location of the Creutz test had been lost over time. Using historical documents, photographs, and engineering records from 1945, the exact location of the test has now been identified (Figure 3.7). In February 2014, a field expedition visited the test location and found historical debris from the Creutz assembly. The artifacts, together with newly declassified photographs, documents, and memoir accounts, now provide greater insight into the preparation activities related to the Trinity test, the uncertainties surrounding the development of implosion technology, and wartime worker experiences.

The P-Site Archaeological Survey Project

In October 2016, large metal objects were discovered at the location of former P-Site, built during the war to support the X-ray and counter method (Figure 3.8). The P-Site buildings had been demolished, and the site was supposedly cleared of any associated Manhattan Project equipment. Archival research using documents, photographs, and engineering drawings eventually identified many of the wartime artifacts. The site's massive detector post, its associated steel base, and the support base for the test's high-explosive charges were among the original Manhattan Project items that had been left on the ground, undisturbed for over 70 years.[39]

3.8. Detector base (left) and detector post (right). *Courtesy of Los Alamos National Laboratory.*

P-Site drawings from 1944 provided clarification about the original location and function of the objects and provided construction and installation information and dimensions. A key historical photograph contributed to the identification of the detector post—a significant Manhattan Project-era artifact—and helped provide an understanding of the role of the equipment and its location at P-Site. Systematic field survey work was conducted in 2017, and additional archaeological features and artifacts from the war years were carefully mapped, including the high-explosives charge platforms used during the tests, a collection of aluminum disks used for special implosion experiments, and the remains of a fireset (a device that supplied energy to the detonators) and cabling used to detonate test charges.

Meaning: The Manhattan Project National Historical Park and Key Interpretive Themes

To aid public understanding of the complex history of the Manhattan Project and its meaning today, the National Park Service has developed interpretive themes, which, according to the park service, are "the key stories or concepts that visitors should understand after visiting a park—they define the most important ideas or concepts communicated to visitors about a park unit."[40] Interpretive themes help to illuminate the historical contexts and present-day meanings and values associated with a specific place, and identify why a particular resource is significant or relevant today.

Interpretive themes have been identified for the Manhattan Project National Historical Park, and topics include the displacement of local populations, worker experiences, the historical context of World War II, and the decision to use the bomb. Additional themes include the role of revolutionary scientific and engineering, the continuing geopolitical legacy of nuclear weapons, and the human costs and environmental consequences of the Manhattan Project.[41] At first glance, the historical meaning of the implosion test sites at Los Alamos seems firmly associated with the scientific history of the development of the Fat Man bomb used against Nagasaki, Japan. By examining this narrative through the lens of other park themes, however, the history of implosion physics can be expanded to provide a greater relevance to people today. Scientists, technicians, military personnel, local Hispanic and Native American workers, and women all supported implosion research and testing. Nobel-prize winning physicists worked side by side with university scientists and graduate students from across the country and across the world. High-explosives implosion tests provided valuable data but also came with heavy human and environmental costs in the form of airborne releases of radioactivity, worker exposures, and the discharge of chemical wastes. A closer examination of one of the implosion test areas, Bayo Canyon Site, reveals a variety of worker experiences and a history of social and environmental impacts that goes well beyond the RaLa program's basic scientific narrative of implosion testing.

RaLa: Worker Stories

The RaLa program, like other implosion testing programs at Los Alamos, was staffed by a diverse wartime work force of technicians and scientists including civilians and military personnel, men and women, and foreign-born and native New Mexicans alike. Workers who supported RaLa research included well-known scientists Bruno Rossi and Luis Alvarez along with lesser known scientific and technical workers like Norma Gross, the WAC chemist who worked on the RaLa chemical

3.9. Norma Gross and Gerhardt Friedlander at Bayo Canyon Site. *Courtesy of Los Alamos National Laboratory.*

separation process (Figure 3.9). Gross worked in Los Alamos's Chemistry and Metallurgy Division during the war.[42] After taking graduate course work at Columbia University in the summer of 1936, Gross received her M.A. degree from Bryn Mawr College in 1938. Her contributions to the Manhattan Project include high vacuum work related to the analysis of carbon, sulfur, and oxygen. Participating in the RaLa experiments in Bayo Canyon, Gross also worked on the chemical separation of radio-lanthanum, partnering with fellow chemist Rod Spence to develop a new precipitation technique for separating it from radiobarium.[43]

RaLa: Human and Environmental Consequences

The RaLa testing area was located in Bayo Canyon, north of the town of Los Alamos, and included two firing sites. Each of the high-explosives assemblies detonated during a firing shot contained a highly radioactive source—radiolanthanum—at its core. Mock pits, standing in for pluto-nium, were made of a variety of metals, including natural and depleted uranium. While slightly radioactive fragments of the test assemblies landed close to the firing sites, higher levels of radioactive contamination were present in the shot clouds that traveled several miles from Bayo Canyon.[44]

More than 250 tests were conducted in support of the RaLa pro-gram from 1944 to the early 1960s.[45] To produce the radiolanthanum, a purification laboratory was set up in the canyon. Hazardous materials associated with this onsite chemistry work included radiobarium and strontium-90.[46] Worker safety at Bayo Canyon Site and the possibility of wind-borne radiation exposures during firing operations were ongoing topics brought up by Los Alamos managers at the time.[47] A weather sta-tion, "Point Weather," was established on a promontory overlooking the test locations to ensure that wind conditions would not create a hazard for the town's residents to the west or for those driving up the nearby main road to Los Alamos during test shots.[48] Ultimately, concern about the proximity of the RaLa test area to the expanding post-war town and laboratory led to its abandonment.[49]

In 1963, the Los Alamos laboratory embarked on a full-scale clean-up of the Bayo Canyon Site. Metal debris associated with the tests was collected.[50] Planned by the laboratory's Health Division, the work was carried out by several groups, including men hired from nearby pueblo communities.[51] A 1963 newsletter reported on the cleanup:

> One of the hardest and most tedious of the many tasks is the picking up of everything 'not native to the area' within a half-mile of two control buildings. The AEC hired 26 men from the Jemez and Zia Indian Pueblos to do the job. During the first ten days, the Indians picked up 20 truckloads of debris which was hauled away and buried.[52]

Site workers walked the area picking up shot debris and collecting metal objects embedded in trees (Figure 3.10). As part of the site cleanup project, some of the uncontaminated firing site buildings were burned, and contaminated buildings were disassembled and buried in place. Concrete bunkers were blown up with dynamite before pieces of the buildings could be removed. Safety procedures were followed according to the standards of the day, and cleanup personnel were instructed to wear leather gloves and to wash their hands before eating. Workers and site equipment were also monitored for radioactive contamination.[53]

After the RaLa firing sites were decommissioned, the area was transferred from the laboratory back to the Atomic Energy Commission (now Department of Energy) in July of 1963. Norris Bradbury ordered a complete assessment of the area before the transfer, and the inspectors reported back that "it was their opinion that the area may be returned to the AEC with no restriction whatsoever as to the use that may be made of it in the future. They said there is no reason to feel that it cannot be used as a picnic ground, for home sites, or any other purpose."[54]

Bayo Canyon Site is now owned by Los Alamos County, and the canyon area is primarily used for hiking and horseback riding. Periodically checked for legacy materials and contamination, the former firing area in the bottom of Bayo Canyon retains little evidence of its wartime use today.[55] However, a small 1.5-acre plot has been identified as a Formerly Utilized Sites Remedial Action Program (FUSRAP) site by the federal government. Soil within the FUSRAP site boundaries is still radioactively contaminated at depths of 8 to 40 feet, a result of RaLa program activities. While the FUSRAP area is safe for recreational uses, it is estimated that the strontium-90 in the buried soil will remain radioactive at unsafe levels until the year 2142.[56] Hailed by Bradbury in 1963 for its importance

3.10. Cleanup workers from Zia and Jemez pueblos. *Courtesy of Los Alamos National Laboratory.*

to Los Alamos' weapons program, the RaLa program also represents the continuing environmental and human health legacy of the Manhattan Project. Not only are buried soils still radioactively contaminated, but predominately Hispanic and Native American communities, located to the north and east of Bayo Canyon Site, were downwind of the RaLa tests, which continued for almost twenty years.[57]

CONCLUSION

The need to see inside the high-explosives implosion at the heart of the Fat Man weapon design was one of the greatest engineering challenges faced by the Manhattan Project. World War II research at Los Alamos resulted in a new field of scientific inquiry, that of implosion physics. Its continuing legacy is reflected in weapons research and stockpile stewardship technology in use during the Cold War and beyond, as illustrated by the PHERMEX and DARHT facilities that were born of early implosion research.

The memory and meaning of implosion science reside both with the archaeological remains at former test sites and with its connections to present-day social themes and environmental and human-health concerns. Interpretive narratives associated with implosion research include the international consequences of the development of nuclear weapons, and, on a more local level, the contributions of all classes of wartime workers—stories that are often undertold and overshadowed. The lasting legacy of radioactive releases and their potential for downwind exposures, exemplified at Los Alamos by the RaLa tests in Bayo Canyon, further drive home the message that the history of implosion science is still relevant today.

Notes

1. Bruno Rossi, *Moments in the Life of a Scientist* (Cambridge: Cambridge University Press, 1990), 75–76; Edwin M. McMillan, "Early Days at Los Alamos," in *Reminiscences of Los Alamos, 1943–1945*, ed. Lawrence Badash, Joseph O. Hirschfelder, and Herbert P. Broida (Dordrecht, Boston, and London: D. Reidel Publishing Company, 1980), 16–17.

2. Lillian Hoddeson, Paul W. Henriksen, Roger A. Meade, and Catherine Westfall, *Critical Assembly: A Technical History of Los Alamos during the Oppenheimer Years, 1943–1945* (New York and Cambridge: Cambridge University Press, 1998), 130, 139.

3. Ibid. 129, 140–146, 279–280; Manhattan Engineer District, "Book VIII Los Alamos Project (Y), Volume 2–Technical," in *Manhattan District History* (U.S. Department of Energy, OpenNet, Circa 1944–1946), VII-30, VII-35.

4. Manhattan Engineer District, VII-30, VII-35; Hoddeson, 144, 279–280.

5. Los Alamos National Laboratory, *Los Alamos: Beginning of an Era, 1943–1945* (Los Alamos, NM: Reprinted by the Los Alamos Historical Society, 1999), 21.

6. Manhattan Engineer District, VII-35; Hoddeson, 245–248.

7. Hoddeson, 130, 139.

8. Hoddeson et al. describes these diagnostic methods in detail on pages 139–156 and 268–281.

9. Manhattan Engineer District, VII-33, VII-35.

10. Hoddeson, 139–156, 268–281.

11. Manhattan Engineer District, VII-33.

12. Rossi, 76.

13. Ibid.

14. Hoddeson, 271–272; The Director–Los Alamos Scientific Laboratory, "Memorandum to The Manager, U.S.A.E.C., Office of Santa Fe Directed Operations; Subject: General Background Data Concerning the Los Alamos Scientific Laboratory, LAB-A-5, September 11, 1947," (U.S. Department of Energy, OpenNet), 7–8.

15. Lilli S. Hornig, "Interview with Dr. Lilli Hornig, June 25, 1986," Transcription of Tape, TR-86-026, Los Alamos National Laboratory Archives, Los Alamos, New Mexico; Wallace Haywood, Dexter McRae, Jonathan Powell, and Betty Harris, *An Assessment of High-Energy Explosives and Metal Contamination in Soil at TA-67 (12), L-Site, and TA-14 (Q-Site), LA-12752-MS* (Los Alamos, NM: Los Alamos National Laboratory, 1995); The Director–Los Alamos Scientific Laboratory, 10.

16. Hoddeson, 155; Manhattan Engineer District, VII-29, VII-30.

17. Gregory S. Cunningham, and Christopher Morris, "The Development of Flash Radiography: The Manhattan Project, PHERMEX and DARHT, Proton Radiography at LANSCE," *Los Alamos Science*, no. 28 (2003).

18. Hoddeson, 278–279; The Director–Los Alamos Scientific Laboratory, 10.

19. L. G. Parratt, J. Allen, and D. P. McMillan, "Memorandum To: J. R. Oppenheimer and R. F. Bacher; Subject: Summary of and Recommendation for the Future of the X-ray Counter Experiment, March 1945," 1–2, Collection A-84-019, Los Alamos National Laboratory Archives, Los Alamos, New Mexico.

20. The Director–Los Alamos Scientific Laboratory, 9.

21. Rossi, 80.

22. Ibid. 82.

23. Ibid. 82, 86.

24. Ibid. 76.

25. Ibid.

26. Ibid. 77.

27. Ibid.

28. Ibid. 82–83.

29. Ibid. 85.

30. U.S. Department of Interior National Park Service, *Foundation Document, Manhattan Project National Historial Park, Tennessee, New Mexico, Washington* (January 2017), 14.

31. F. G. Gosling, *The Manhattan Project: Making the Atomic Bomb* (Washington, DC: U.S. Department of Energy, 2001), 55.

32. Cunningham, "The Development of Flash Radiography: The Manhattan Project, PHERMEX and DARHT, Proton Radiography at LANSCE."

33. "Text of Dr. Bradbury's News Conference," *LASL News*, August 1, 1963, 6.

34. Los Alamos National Laboratory, "All Innovations: 70 Years of Innovations," (2017). http://www.lanl.gov/science-innovation/features/innovations/all-innovations.php (accessed September 2017).

35. Charles L. Mader, Timothy R. Neal, and Richard D. Dick, ed. *LASL PHERMEX Data, Volume 1*, Los Alamos Series on Dynamic Material Properties (Berkeley: University of California Press, 1980), 1.

36. Los Alamos National Laboratory, "All Innovations: 70 Years of Innovations."

37. Edward C. Creutz, Martyn H. Foss, and Rolf E. Peterson, "Full-Scale Implosion at Pajarito, LA Report 346 (unclassifed notes), August 10, 1945," Los Alamos National Laboratory Archives, Los Alamos, New Mexico; G. B. Kistiakowsky, "Trinity—A Reminiscence," *Bulletin of the Atomic Scientists* 36, no. 6 (June 1980).

38. Kistiakowsky, "Trinity—A Reminiscence."

39. Parratt, 1–2; Hoddeson, 277–279; The Director–Los Alamos Scientific Laboratory, 10.

40. National Park Service, 26.

41. Ibid.

42. Dorothy McKibbin, "McKibbin Cards," Index cards with early Los Alamos personnel information, Los Alamos National Laboratory Archives, Los Alamos, New Mexico.

43. "CM Division," Box 14, File 8, Collection A-84-019, Los Alamos National Laboratory Archives, Los Alamos, New Mexico; "Questionnaires, Ca. 6/45–7/45," Collection A-84-019, Los Alamos National Laboratory Archives, Los Alamos, New Mexico; Hoddeson, 326. Note: A typographical error in Hoddeson et al. on page 326 incorrectly attributes the RaLa chemistry work to "Norman" Gross.

44. U.S. Department of Energy, *Phase 1: Installation Assessment, Los Alamos National Laboratory, Comprehensive Environmental Assessment and Response Program [CEARP] (Working Draft)* (Albuquerque Operations Office, Albuquerque, New Mexico. Los Alamos National Laboratory, Los Alamos, New Mexico, 1986), TA-10-3; "Bye Bye Bayo Site," *LASL News*, May 23, 1963, 7.

45. U.S. Department of Energy, TA-10-1.

46. Ibid. TA-10-4.

47. J. E. Dummer, J. C. Taschner, and C. C. Courtright, *The Bayo Canyon/Radioactive Lanthanum (RaLa) Program, LA-13044-H* (Los Alamos, New Mexico: Los Alamos National Laboratory, 1996), 12, 16–22.

48. Roger W. Ferenbaugh, Thomas E. Buhl, Alan K. Stoker, and Wayne R. Hansen, *Environmental Analysis of the Bayo Canyon (TA-10) Site, Los Alamos, New Mexico* (Los Alamos, New Mexico: Los Alamos National Laboratory, 1982), 6, LA-9252-MS; Dummer, 12.

49. "Bye Bye Bayo Site," 7–8; The Director–Los Alamos Scientific Laboratory, 9.

50. U.S. Department of Energy, TA-10-3.

51. "Bye Bye Bayo Site," 8–10.

52. Ibid.

53. Ibid. 10.

54. "Laboratory Returns Bayo to AEC," *LASL News*, August 1, 1963, 16.

55. U.S. Department of Energy, TA-10-1.

56. Darina Castillo, PhD, General Engineer, U.S. Department of Energy, Office of Legacy Management, personal communication, June 1, 2017.

57. "Laboratory Returns Bayo to AEC," 16.

CHAPTER FOUR

Herbert M. Parker, Health Physics, and Hanford

Ronald L. Kathren

INTRODUCTION

The closing years of the nineteenth century were marked by two extraordinary scientific discoveries that excited the world and radically changed the direction and indeed the understanding of physics and the structure of the atom.[1] The first was the discovery of X-rays on November 8, 1895, by German physics professor Wilhelm Konrad Roentgen, which was reported to the world in the first few days of January, and was so electrifying that many experimenters, including Thomas Edison, began their own studies the very day they learned of the discovery. Within a matter of days, X-ray pictures of the body had been applied to medical diagnosis and to various commercial endeavors, mostly to photographic or fluoroscopic examinations of the hand of a customer. Only two months later, on March 2, 1896, French physicist Antoine Henri Becquerel reported that, much like X-rays, unknown invisible emanations from a uranium-rich mineral could blacken a photographic plate from which light had been totally excluded and could also cause a charged electroscope to discharge. Thus, the discovery of radioactivity. The mysterious emanations from uranium were similar to X-rays, and later studies revealed a whole host of new elements and the apparent transmutation from one element to another, like the philosopher's stone of the Middle Ages which was supposed to turn base metals like lead into gold. Then, on December 21, 1898, Marie and Pierre Curie discovered radium, far more radioactive than uranium, which quickly found application in self-luminous paint and other products, and as a patent medicine tonic. For the next 30 years radium was hailed as a panacea and tonic, and applied (largely unsuccessfully) to treatment of all sorts of medical conditions by legitimate medical practitioners. It also found its way into numerous nostrums and patent medicines, sometimes with disastrous health effects.

But a sinister downside soon tempered the excitement generated by the two discoveries. Within a matter of weeks after being put into use, reports of possible X-ray injuries and skin burns began to surface. While this observation ultimately led to the use of X-rays and later radium to destroy malignancies and skin lesions, initially it was not generally accepted that the injuries observed were in fact attributable to X-rays. Indeed, it took almost a full decade before there was general consensus of the potential harmful effects of X-ray overexposure. Another decade passed before the first organized efforts at radiation protection appeared, some two decades after the discoveries, beginning with the recommendations of the British Roentgen Society (1915) and protection rules put forth by the American Roentgen Ray Society (1922). Initial protection efforts were slow and spotty. It was not until 1924 that the first exposure dose limit was proposed by American physicist Arthur Mutscheller.

In that same year, American dentist Theodore Blum identified a disease called "radium jaw" in his patients, largely young women, who painted clock and watch dials with radium containing radio-luminescent paints. Following this initial observation, originally reported as a footnote to Blum's article in the *Journal of the American Dental Association,* came reports of other dial painters who were sickened and died from osteogenic sarcomas. Spurred by the disaster of the radium dial painters and injuries associated with the misuse of radiation in medicine, and further buttressed by the subsequent effort to build the atomic bomb, radiation protection became a serious concern, ultimately resulting in the establishment of two new and somewhat overlapping board certified professions: *medical* (also known as *radiological*) *physics,* primarily concerned with safe applications of radiations of all types to the diagnosis and treatment of disease, and *health physics,* which dealt with the protection of people and the environment from potentially harmful effects while at the same time realizing the beneficial applications. This essay briefly outlines the seminal contributions made by Herbert Parker to both of these nascent

4.1 Herbert Parker in 1938, around the time he emigrated from England to the United States.

professions, with special emphasis on his contributions to the exceptional radiological safety record compiled at the Hanford Site, where all manner of new and unknown radiological hazards and especially those associated with plutonium were present during World War II and beyond.

Herbert Myers Parker was born on April 13, 1910, to William Henry and Elizabeth Parker in the small English city of Accrington, situated in Lancashire about 20 miles north of the city of Manchester. At the time of his birth, Accrington was largely a working-class city with a population of about 45,000. The primary industries in the area related to textile mills, coal mining, and engineering activities associated with industrialization. At an early age, Parker demonstrated strong intellectual ability that brought him recognition, including being named Manchester Ship Canal Scholar in 1927. He enrolled to read physics at the University of Manchester— then, as now, a world-class university—which awarded him two important scientific prizes during his undergraduate years: the Higginbotham Prize in 1928 and the H. G. Mosely Prize in 1929. In 1930, he received his B.Sc. in physics, and was awarded an M.Sc. the following year. His thesis research was x-ray analysis of iron pyrites using Fourier analysis, and was of sufficient import and quality to merit publication in *Philosophical Magazine*, a major peer reviewed scientific journal. His desire was to continue his education to obtain a Ph.D. and pursue an academic career, but the worldwide depression of the 1930s left him without the financial means to do so. Fortuitously, he was recruited by the preeminent British radiologist James Ralston Kennedy Paterson, joining the staff of the Holt Radium Institute of Christie Hospital in 1932, where he began a professional career in medical physics.

THE MANCHESTER YEARS AND EARLY PHYSICS CONTRIBUTIONS (1932-1943)

In 1932, Ra-226 was for all practical purposes the only available radionuclide, and its use was largely confined to medical applications, radio-luminescent paints, and similar products. Medical applications were primarily the treatment of malignancies, as well as uterine and cervical cancers, which had been pioneered in 1903 by Margaret Cleaves. Three decades later, there was still no commonly accepted radium treatment methodology or a unified system of dosimetry. At Holt, Parker was to collaborate with Paterson on development of a unified system of radium therapy for cervical and uterine cancers, which led to the development of a broader dose-based system of cancer therapy known as the Paterson-Parker or Manchester System.[2] The Paterson-Parker System provided a practical standardized method of achieving homogeneous reproducible doses to a broad range of malignancies of various types, sizes, and shapes, using arrays of radium needles designed to maximize the tumor dose and minimize the dose to healthy surrounding tissues. As such, it represented a major advance in the treatment of malignancies.

Parker's major contribution was what can best be described as the mathematically elegant calculation of doses from various geometries of radium needles, including a series of arrays and techniques for molds for intraoral and superficial therapy of all types. To accomplish this, he used sources of unequal intensity and devised a number of complex but easily made geometric arrangements that could be used for external therapy as well. The method has been reprinted and updated several times in book form and the basic techniques are still in use today with various radionuclides, some eight decades later. Subsequently, in conjunction with colleague W. J. Meredith he developed an apparatus for the exact three-dimensional reconstruction of the arrangement of needles used for a radium implant, which provided improved accuracy to the Manchester System.[3] Over the years, the Paterson-Parker techniques have been identi-fied as one of most significant developments in radiological physics, hav-ing been ranked with the discovery of artificial radioactivity and Parker compared favorably with Enrico Fermi.[4] Other activities at Holt were largely related to improving dose measurements and included developing improved ion chamber monitoring instrumentation for X-rays, which were also used for radiotherapy. This included the development of small ion chambers with sufficient sensitivity to measure small X-ray doses and a response to various X-ray energy spectra that was similar to that of soft tissue and hence in a sense the forerunner of modern tissue equivalent detectors. He studied the effects of backscatter on X-ray and deep doses, which because of attenuation in the overlying tissue were often much lower than at the surface point of entry.[5]

In September 1937, Parker attended the Fifth International Congress of Radiology held in Chicago, where he presented a paper entitled "Dosage Measurements by Simple Computations."[6] This was, he privately noted, the "[f]irst paper I read in the US."[7] The following year he emigrated to the United States to accept an invitation to pursue his interests in deep voltage X-ray therapy at the Swedish Hospital Tumor Institute in Seattle, working with physician Simeon Theodore Cantril. At Swedish he was placed in charge of radiological physics and embarked on an energetic research program of what were then known as supervoltage X-rays, X-rays with higher energies than those produced by conventional apparatus and hence were of greater value in the treatment of deep-seated tumors. Swedish became a world leader in treating cancers with supervoltage X-rays, and Parker ultimately coauthored a book with Cantril and Franz Buschke entitled *Supervoltage Roentgentherapy,* that was identified by radiologist-historian E. R. N. Grigg, in his monumental history of radiology, as a "Memorable Book."[8] While at Swedish, Parker also became concerned with radiation overexposures and to that end coauthored a paper about the hazards of repeated fluoroscopies in infants that was published in 1942 in the *Journal of Pediatrics.*[9] This remarkably prescient article predated by more than a decade the epidemiologic studies of Alice Stewart and coworkers that first revealed an increased incidence of childhood leukemia in children irradiated *in utero* when compared with unirradiated controls.[10]

HANFORD: THE WAR YEARS

Prior to World War II, there were no health physicists, and health physics as a profession did not exist. Radiation safety was not a priority and was of secondary concern among users of X-rays and radium. But the establishment of the Manhattan Project to build an atomic bomb changed all that. Even before construction began on the Hanford facility, there was contemplation and concern about the health and environmental aspects of radium, and especially plutonium production at Hanford, spurred by the tragedy of the radium dial painters. General Leslie Groves, in his memoirs of the Manhattan Project, specifically noted that in the design of the plutonium production plant, "all possible care was taken to safeguard the health of people who would be working in them."[11] As to the environment, General Groves was cautioned by his superior officer General T. M. Robins, who told him, "[w]hatever you may accomplish, you will incur the everlasting enmity of the entire Northwest if you harm a single scale on a single salmon."[12]

In 1942, Cantril, with whom Parker had worked closely since arriving in the United States, left the Swedish Hospital Tumor Institute in Seattle to join the Manhattan Project, and was initially appointed as director of clinical medicine at the facility that was then known as the Metallurgical Laboratory, or MetLab. He urged his former colleague Herb Parker, by now a prominent scientist in his own right, to join him in Chicago. This Parker did in late 1942, and, despite being a foreign national, was named head of Physical Measurements Section, concerned with developing instrumentation to detect and measure radiation levels for the protection of workers. His arrival at the MetLab was noted by no less a personage than Glenn Seaborg in his journal on January 19, 1943, thusly: "An expert on radiation health problems has started on the Metallurgical Project. He is Herbert M. Parker."[13] Shortly thereafter, Cantril was appointed health director for Clinton Laboratories near Oak Ridge, Tennessee, and Parker joined him there as chief of the Health Physics Section, where he established the radiation safety program for the first of the major American nuclear energy research and production facilities. Much of the initial research on plutonium was carried out at Clinton Labs, research that would yield invaluable results for the then under-construction Hanford plutonium production facility. The root origin of the term "health physics" lies in the need and desire for secrecy within the Manhattan Project; the term conveyed nothing.[14] Since Parker's group were largely physicists working on problems related to health, the term was obscure and not to be found in any dictionary of the time. Parker was one of the first three persons to bear the title, which he personally deplored, feeling that the proper title should be more representative of the tie to medicine, hence medical or radiological physics.

As the construction of the Hanford Site moved rapidly forward, there were concerns over the health hazards of plutonium which were then largely unknown. Accordingly, a large and comprehensive medical program reporting to the plant director and coordinated under the oversight of Robert S. Stone, a physician who served as associate director for health of the Manhattan Project, was initiated. The three site programs had in common a medical organization and a health physics organization and also carried out site specific research.[15] On the recommendation of Nobel Laureate and Metallurgical Laboratory Director Arthur Holly Compton, Parker was selected to head the radiation safety program at Hanford. Compton, in a letter to Hanford contractor E. I. DuPont de Nemours, had noted that Parker was the best man for the job.[16] And so Parker came to Hanford in August 1944—about two months before the first plutonium production reactor went critical—as head of the Health Physics Section of the medical division headed by his old colleague Cantril, who was appointed medical superintendent.

At Hanford, Parker established a comprehensive operational radiation protection program, including a system of exposure dose limits, medical testing, instrument development, and personnel monitoring for thousands of workers. Many of the operational techniques used to limit exposure and ensure worker protection against radiation exposure remain a part of the armamentarium of professional health physicists to this day. This program was fully integrated with the medical monitoring performed by the medical section and was based on the then accepted concept of a tolerance dose; that is, a level of exposure that could be tolerated day in and day out over a lifetime with no demonstrable ill effect. In collaboration with Cantril, Parker prepared a 33-page classified report entitled "The Tolerance Dose," summarizing with great clarity what was known about biological effects of radiation and how this knowledge applied to protection of people.[17] This report described how the conservative radiation exposure limits used in the Manhattan Project had been established. In addition to adhering to these exposure limits, investigations of working conditions were performed when an employee exposure exceeded half of the already conservative exposure limits. It is a tribute to Parker and his colleagues that no employee at Hanford during the war years received an exposure to radiation that was observed to be frankly injurious, and of course there were no deaths attributable to radiation exposure. Tens of thousands of physical examinations and routine blood tests were performed, along with individual monitoring of the radiation dose to workers via pocket ionization chambers coupled with preemployment physical examinations to screen out workers whose fitness might be questionable.[18]

Parker's single most outstanding contribution, perhaps—and certainly the most lasting one during his Manhattan Project days—related to radiation dose quantities and the units used to describe their magnitude. Recognizing that the only measurement unit for ionizing radiation then in use, the *roentgen*, was inadequate to properly characterize doses from all types of radiations to which workers might be occupationally exposed, in 1948, Parker, at a meeting of the Radiological Society of North America, defined two new radiation quantities and units to better characterize and control personnel exposures.[19] These were integrated with the roentgen, which he retained but limited to measurement of ionization produced in air to quantify exposures from X- and gamma-rays under specified conditions. For absorbed dose from any type of ionizing radiation he defined a new unit he called the *rep*, an acronym created from *roentgen equivalent physical*, and which later was named "the parker" in his honor.[20] And, to take into account the varying biological effectiveness of different kinds of radiations, he defined a unit for biological dose measured in units of *rem*, an acronym derived from *roentgen equivalent man*.

Now, three quarters of a century later, this insightful system of dose units, modified and expanded to include subsequent knowledge, remains the basis of radiation protection exposure standards, and the rep and the rem are the direct ancestors of the modern radiological SI units *gray* and *sievert* which apply to absorbed dose and effective (biological) dose, respectively. Parker also introduced the concept of maximum permissible concentration (MPC) for inhaled radioactivity in air, and applied it to plutonium, an alpha emitter like radium based on the rem dose to the lung from inhaled plutonium. Despite numerous scientific advances relating to the biokinetics of plutonium in lung and weighting factors to account for alpha radiation, the original MPC he calculated in 1944 is approximately the same as the permissible level today, a tribute to his prescience.

Parker remained concerned with environmental impacts and disposal of long-lived radioactive wastes. He was uneasy with potential adverse impacts of releases of radioactivity into the environment, whether from the reactor cooling water discharges to the river or the gaseous stack releases that occurred from the processing of the irradiated fuel slugs to remove the plutonium, an operation done remotely from within a heavily shielded facility. He thus evaluated the release of particulate fission product radioactivity. To his credit, he foresaw the potential hazards of radioiodine releases to the environment, recognizing that the radioiodine thus released, if inhaled or ingested by people, would then be concentrated in the thyroid and deliver a proportionately large dose to that tiny organ.

Wartime exigencies and the need for haste in developing the atomic bomb to bring the war to an early end were paramount and dictated that irradiated fuel, which contained I-131, be processed quickly after removal from the reactor. Balanced against this was the need to delay the processing to allow the I-131 to naturally decay or "cool" and hence reduce the levels that would be released, but also for minimizing contamination of vegetation which could affect foodstuffs or the thyroids of animals, including cows and goats whose milk would be consumed by humans and would irradiate the thyroid from the radioiodine it contained. Parker thus fought hard for an increase in the holding time of irradiated fuel from 60 to 90 days before reprocessing, to reduce the amount of I-131, with a half-life of 8.05 days, released from the stacks by 99 percent through radioactive decay. It was a battle that lasted beyond the end of the war, but Parker finally emerged as the winner. And, his concern was not only I-131, but all the radionuclides, particulate and gaseous, in the releases, and the doses and potential biological effects to the natural fauna and flora from waste discharges to the environment.

Postwar Laboratory Years: General Electric (1947-64) and Battelle (1965-71)

On January 1, 1947, General Electric Corporation (GE) assumed the responsibility for the management of the Hanford Site from the du Pont Corporation. Under the new contractor, Parker initially continued on in his dual role as director of the Health Instrument Division and assistant superintendent of the Medical Department. Following a brief stint as consultant to the director, Parker was named director of the new Radiological Sciences Department, whose responsibilities included health physics, radiation biology, and environmental radioactivity research. In 1955 he rose to the position of director of the newly created (and independent) Hanford Laboratories, which quickly achieved worldwide recognition for their research activities. Ten years later, the Hanford Site activities were diversified, and Battelle Memorial Institute assumed the responsibilities for what had been the domain of the Hanford Laboratories. It also took on a number of other research-related activities largely tied to the development of nuclear power under a contract to operate the Pacific Northwest (now National) Laboratory of the AEC, as well as performing other related governmental and private research. Parker stayed on with Battelle as associate director and special consultant to the director until his retirement in 1971.

During the postwar years, and especially the early GE years, Parker was to make many significant technical contributions and broaden his scope beyond occupational radiation safety, despite heavy administrative responsibilities. He built the Hanford Laboratories into a formidable world-class research institution, gaining an international reputation in the process. Under Parker's leadership, the laboratories carried out research in diverse areas, largely concerned with safe commercial applications of nuclear energy, affording Parker an unparalleled opportunity to pursue his personal interests relating to health, safety, and environment. Despite heavy management responsibilities, he was never far removed from the day-to-day conduct of radiobiological and health physics research, guiding, contributing ideas, and otherwise facilitating the research in this area.

He organized and built a world-class radiobiology research program, later teaming with William J. Bair, who became the head of the radiobiology department. (Bair himself would go on to achieve international recognition as a radiation biologist, with special expertise in inhalation toxicology, lung dynamics, and biological aspects of plutonium and other radioactive substances.) So successful were the laboratories that they were the subject of an article in *Business Week* for May 6, 1961, an issue which featured a photo of Herb Parker on the cover.

Under GE, Parker continued the studies he had begun during the war years as well as expanding broadly into more diverse safety related areas. Continued evaluation of radioactivity releases was among the problems he tackled first, which led to methods and recommendations to reduce particulate emissions and radioiodines based on dose limitations to

4.3. Herbert Parker on the cover of the May 6, 1961, issue of *Business Week* magazine, standing in front of a small experimental test reactor. Note the containment building.

4.2. Herbert Parker (left) and William J. Bair at a Hanford Life Sciences Symposium Conference circa 1975.

members of the public. He identified the importance of the vegetation-cow-milk-human thyroid pathway for radioiodine, which later assumed great importance in establishing radioiodine release limits from nuclear power plants sufficient to protect the general public and especially children who typically were large milk consumers. His interest in limiting radioactivity releases to the environment led to an important classified memorandum evaluating the potential problems and suggesting mitigatory actions written in 1947.

The following year, he published a lengthy article entitled "Health Physics, Instrumentation and Radiation Protection" in Volume 1 of *Advances in Biological and Medical Physics.*[21] This article was essentially the first textbook of operational health physics. It included consideration of the underlying scientific bases, control techniques, and instrumentation from an applied standpoint, and as such was a *vade mecum* for those engaged in radiation safety. The historical significance of this article is demonstrated by the fact that it was one of 22 seminal papers selected for reprint in the June 1980 silver anniversary issue of the journal *Health Physics,* more than three decades after its publication. Also selected as one of the 22 was the system of dose quantities and units Parker had developed and used at Hanford, and which became and remains a fundamental basis of radiological protection practice.[22] The importance of Parker's two contributions is underscored by the fact that he was the only person to have two items selected for inclusion in this special commemorative issue.

During his tenure as head of the GE Hanford Laboratories, Parker broadly expanded his areas of interest from occupational radiation safety to more far-reaching concerns relative to applications of nuclear technology generally, and to the development and safe use of civilian nuclear power in particular. Parker wrote a number of influential landmark papers relating to reactor safety, nuclear waste disposal, and radiological concerns that were harbingers of the future, several of which were presented in July 1956, at the first United Nations Conference on Peaceful Uses of Atomic Energy in Geneva. One was a brief historical note describing radiation protection operations at Hanford, which had built and operated large plutonium production reactors and chemical separations of plutonium.[23] A second paper dealt with environmental hazards,[24] while a third, coauthored with his longtime friend and colleague John W. Healy, was a prescient prospective examination of the environmental consequences of a serious nuclear reactor accident to justify the use of reactor containment structures, a recommendation that has been proven out in the few serious reactor accidents that have occurred over the subsequent six decades.[25]

In addition to his strictly scientific or technical contributions, Parker profoundly influenced the development of radiation protection policy both nationally and internationally. In the United States, hearings were held by Congress immediately after the surrender of Japan, and on August 1, 1946, an act creating the Atomic Energy Commission (AEC), calling for civilian control and management of nuclear energy, was signed into law by President Harry Truman. Spurred by the scientific papers of Parker, among others—and by visions of widespread potential beneficial uses of radiation in medicine, research, and various commercial and military applications—efforts began in the United States, Britain, Canada, and within the United Nations, to develop policies for control and management of nuclear energy, including broad-based radiation protection policy.

Parker was heavily involved in the development of such policies, serving on or otherwise contributing to various working groups producing significant and important reports such as those of the United Nations Scientific Committee on the Effects of Atomic Radiation (UNSCEAR) and EURATOM, via his membership on the NCRP, and through a small but significant number of publications in the open peer reviewed literature. In May 1960, he testified on the basis and application of radiation protection criteria before the Joint Committee on Atomic Energy (JCAE) of the U.S. Congress, identifying five specific problems related to setting and applying adequate radiation protection standards. Primary among these was insufficient quantitative knowledge of the deleterious effects of radiation to fully support the adoption of the risk principle as the basis for protection limits, a problem that still plagues radiation protection standards-setters to this day. Another was his concern with potential future widespread use of radiation with concomitant low-level exposure to members of the general population, and the degree of enforcement necessary to ensure adequate safety. Two years later, he again testified before the JCAE hearing "Radiation Standards, Including Fallout," noting in his summary the lack of any outstanding or basic change regarding radiation protection subsequent to his earlier testimony. At the conclusion of his testimony, JCAE hearing chairman Congressman Melvin Price commended Parker for expanding and bringing up to date the knowledge of the committee.

In 1946, the NCRP, which had been dormant during the war, recognized the need for action with respect to radiation protection and underwent a major reorganization and expansion. Parker was named a member of the Executive Council, and one of only seven individuals and chairman of what would become NCRP Scientific Committee 1, Basic Radiation Protection Criteria. It was a role he served in for several decades, strongly influencing the deliberations and recommendations of the committee through his scientific knowledge and contributions, insights, judgement, and prescience. In his role with the NRCP, he was in a position to influence not only scientific matters, but also radiation protection policy. He played a key role in this regard as chair and major contributor to NCRP Report 39, "Basic Radiation Protection Criteria" which provided a succinct and complete review of the scientific bases of radiation protection, and was an important influence in the establishment of radiation protection standards. Although published nearly 50 years ago,

NCRP 39 is still a valuable resource for those engaged in the practice of health physics. For his numerous influential and often groundbreaking contributions, the NCRP honored Parker by naming him the first Lauriston S. Taylor Lecturer when the lectureship was established in 1977.[26] As noted above, from his earliest days at Hanford, Parker recognized the fundamental importance of controlling the releases of radioactivity into the environment from a radiological safety standpoint, anticipating the implications that would be brought about by nuclear power plants and radioactive releases and wastes associated with their operation. More significantly, he understood that knowledge gained through research would be fundamental to devising limits for releases not only from nuclear power plants but also from uses of radioactivity generally, whether from research, medical, or industrial applications. To that end, he strongly encouraged and supported research of environmental monitoring of radioactivity releases and promoted research to better understand and determine atmospheric diffusion and fallout patterns from releases of radioactivity into the environment. Thus, instrumented towers were used at Hanford to measure atmospheric diffusion parameters and, along with monitoring results, develop empirical mathematical models to predict airborne concentration and deposition levels and patterns.

Radioactive waste management was one of Parker's most long-standing concerns, dating from his earliest days at Hanford. Recognizing what the future would hold, he encouraged others to get involved in this neglected area of research. He realized that the advent of nuclear power and widespread use of radioactive materials would, of necessity, require means to manage radioactive wastes, including its appropriate disposal or isolation to ensure the safety of both humans and the environment. Accordingly, in 1948 he drafted a ten-page classified report dealing with the disposal of Hanford wastes into the soil, concluding that the present waste disposal practices would likely be adequate for the next 50 years. This early report also included the caveat that substantial changes in waste volumes or in the physicochemical status of the waste could alter this conclusion, and recommended the unresolved waste problems be "managed by a first-class soil chemist."[27]

A year and a half later, he published a more comprehensive but still preliminary 57-page classified report dated August 2, 1949 (declassified

February 1, 1974 as AEC Report HW-14058) entitled "Hazards of Waste Storage at Hanford Works." Only 20 copies were printed; the first 10 of these went to the AEC and the remainder to onsite distribution and file. This report did not reconsider the tank storage situation but instead focused on radioactive waste storage in a reactor and uncontrolled releases to the environment. It painted a rather grim picture of potential environmental contamination consequences, including potential biological impacts not only on people but on the fauna and flora. While by contemporary standards the specific assumptions and calculational results may appear simplistic and wanting, the methodology and reasoning continue to influence the development of the current highly detailed approach that is the hallmark of modern environmental impact studies.

AFTER HANFORD

In 1971, Parker retired from active employment on the Hanford Site and established his own consulting company, HMP Associates. He continued to serve on numerous boards and committees, including membership on the NCRP, and as a consultant to the congressionally mandated U.S. Advisory Committee on Reactor Safeguards and its spinoff sister the U.S. Advisory Committee on Nuclear Waste, and as a member of the Scientific Advisory Committee of the U.S. Transuranium and Uranium Registries. His concern for the aftermath of a serious accident involving a nuclear power reactor led him to join with medical and health physics colleagues to form Radiation Management Corporation, a company with specific expertise in management of radiation related accidents. And, of course, he continued to serve informally as a behind-the-scenes generally unrecognized *ad hoc* consultant to committees and individuals who sought his assistance.

The many important fundamental and lasting contributions to the emerging profession of health physics led to recognition via a number of professional awards and honors over the years. Early in his career Parker was elected a fellow of the Institute of Physics in Great Britain and ultimately to fellowship in no less than six scientific organizations. He later attained board certification in Radiologic Physics from the American Board of Radiology and in Health Physics from the American Board of Health Physics. In 1955, he was named Janeway Lecturer and

Medalist by the American Radium Society, an award given annually to a single individual in recognition of scientific contributions. In 1971, he was the recipient of the Health Physics Society Distinguished Scientific Achievement Award and selected in 1977 as the inaugural Lauriston S. Taylor Lecturer of the NCRP. Other honors included election to the National Academy of Engineering (1978) and the American Association of Physicists in Medicine William D. Coolidge Award (1979). He also served a term on the boards of directors of the American Nuclear Society and the Health Physics Society. There is a bit of irony with respect to the latter, as Parker, with his medical physics background, felt strongly that the newly established profession of health physics should ally with the medical physicists. And so, when the Health Physics Society held its organizing meeting in 1955, Parker was not among the attendees, and indeed worked toward a nexus between the health physics and medical communities. But this was not to be, and ultimately Parker joined the Health Physics Society, serving a full term on the board of directors (1971-1974).

Despite his seminal contributions to the science and art of radiation safety, Parker had many other interests. His hobbies included contract bridge and stamp collecting. An accomplished ballroom dancer, Parker enjoyed demonstrating this skill at dinners and social events, and was a gracious and charming host at social gatherings at his home. He and his wife established a large garden on their small acreage bordering the Columbia River, achieving international renown for growing world-class irises of all colors and varieties. He passed away on March 5, 1984, in Richland, the small city in eastern Washington state that had been his home for four decades.

It seems fitting to conclude this essay with Merril Eisenbud's testament to Parker, written as a memorial for the National Academy of Sciences: "His many colleagues throughout the world are grateful for his accomplishments and for the privilege of having been his associates and friends," Eisenbud wrote. "The extraordinary safety record of the atomic energy industry in the United States and elsewhere is the result, to a large degree, of the fundamental pioneering work of Herbert Parker."[28]

Notes

1. This essay is an expanded written version of the oral presentation by the same name given at the "Legacies of the Manhattan Project" Conference held at Washington State University at Tri-Cities, March 16, 2017. In preparing this written version, the author has drawn freely from the book *Herbert M. Parker: Publications and Other Contributions to Radiological and Health Physics*, R. L. Kathren, R. W. Baalman, and W. J. Bair, eds. (Columbus, OH: Battelle Press, 1986) as well as his own personal knowledge and experiences with Herbert Parker.

2. See H. M. Parker, "A Dosage System for Gamma-ray Therapy, Part II," *British Journal of Radiology* 7 (1934): 612–632. See also H. M. Parker, "A System of Dosage for Cylindrical Distributions of Radium, Part II: An Outline of the Physical Basis of the Dosage System," *British Journal of Radiology* IX (1936): 494–499.

3. H. M. Parker and W. J. Meredith, "A Radium Implant Reconstructor," *British Journal of Radiology* XII (1939): 499–503.

4. O. Glasser, E. H. Quimby, L. S. Taylor, and J. L. Weatherwax, *Physical Foundations of Radiology* (New York: Paul B. Hoeber, Inc, 1952) 14.

5. H. M. Parker, "The Dependence of Back-Scattering," *Acta Radiologica* XVI (1935): 705–715.

6. H. M. Parker, "Dosage Measurements by Simple Computations," *Radiology* 32 (1939): 591–597.

7. *Herbert M. Parker*, 97.

8. E. R. N. Grigg, *The Trail of the Invisible Light* (Springfield, IL: Charles C. Thomas, 1965), 839.

9. F. Buschke and H. M. Parker, "Possible Hazards of Repeated Fluoroscopies in Infants," *The Journal of Pediatrics* 21 (1942): 521–533.

10. A. M. Stewart, J. W. Webb, B. D. Giles, and D. Hewitt, "Preliminary Communication: Malignant Disease in Childhood and Diagnostic Irradiation In-Utero," *Lancet* 2 (1956): 447–451.

11. Leslie R. Groves, *Now It Can Be Told: The Story of the Manhattan Project* (New York: Harper and Brothers, 1962), 86.

12. Ibid.

13. *The Plutonium Story: The Journals of Professor Glenn T. Seaborg 1939–1946*, R. L. Kathren, J. B. Gough, and G. T. Benefield, eds. (Columbus, OH: Battelle Press, 1994), 234.

14. B. C. Hacker, *The Dragon's Tail: Radiation Safety in the Manhattan Project, 1942–1946* (Berkeley: University of California Press, 1987), 31.

15. J. E. Wirth, "Medical Services of the Plutonium Project," in *Industrial Medicine on the Plutonium Project*, R. S. Stone, ed. (New York: McGraw Hill, 1951), 19–35.

16. *The Plutonium Story*, 395–96.

17. S. T. Cantril and H. M. Parker, "The Tolerance Dose," United States Atomic Energy Commission Report MDDC-1100 dated January 5, 1945; declassified June 30, 1947.

18. See S. T. Cantril, "Industrial Medical Program – Hanford Engineer Works," and S. T. Cantril and H. M. Parker, "Status of Health and Protection at the Hanford Engineering Works," both in *Industrial Medicine on the Plutonium Project*, 289–307 and 476–484.

19. H. M. Parker, "Tentative Dose Units for Mixed Radiations," *Radiology* 54 (1950): 257–261.

20. To old timers familiar with Parker's contribution, the "p" in rep was respectfully and humorously said to stand for Parker.

21. H. M. Parker, "Health Physics, Instrumentation and Radiation Protection," *Advances in Biological and Medical Physics*, 1 (1948): 226–282.

22. "Tentative Dose Units."

23. H. M. Parker, "Radiation Exposure Experience in a Major Atomic Energy Facility," in *United Nations International Conference on the Peaceful Uses of Atomic Energy*, Geneva, Switzerland, 1955–8, Volume 1 (New York: United Nations, 1956), 266–269.

24. H. M. Parker, "Radiation Exposure from Environmental Hazards," in ibid., 305–310.

25. H. M. Parker and J. W. Healy, "Environmental Effects of a Major Reactor Disaster," in ibid., 106–110. Notably, American nuclear power reactors have all been enclosed within containment vessels. Others, such as the large power reactor at Chernobyl in the Ukraine, did not have a containment vessel to prevent the escape of very large quantities of fission products into the environment following the catastrophic accident there in 1986.

26. Parker's lecture was published as "The Squares of the Natural Numbers in Radiation Protection," *Lauriston S. Taylor Lecture Series in Radiation Protection and Measurements Lecture No. 1*, (Washington: National Council on Radiation Protection and Measurement, 1977).

27. H. M. Parker, "Speculations on Long-Range Waste Disposal Hazards," U.S. Atomic Energy Commission report HW-8674 dated January 26, 1948. Declassified June 24, 1986.

28. M. Eisenbud, "Herbert M. Parker," *Memorial Tributes: National Academy of Engineering* Volume 3 (1989): 378–383.

CHAPTER FIVE

"The Atom Goes to College"
The Teaching Reactors that Trained the Atomic Age

David P. D. Munns

The reactor core began to glow blue around 4:00 in the morning. The agonizingly slow process of making the reaction go critical had been going on since early afternoon the previous day. After waiting around for hours with nothing to see or say, University of Michigan president Harlon Hatcher and his wife left about midnight. Guided by notable physicist Ralph Sawyer and nuclear engineer Henry Gomberg, the reactor staff and nuclear engineers stayed, however, until the newest nuclear reactor in the United States went critical. Theirs was not a reactor for weapons or power, however. It was not attached to the Atomic Energy Commission or the United States military. Instead, built by the Ford Motor Company, the facility belonged to the Michigan Memorial Phoenix Project and the Nuclear Engineering department of the University of Michigan. It was a reactor for education. Known as the Ford Nuclear Reactor, it taught a generation how to build and operate nuclear reactors.

What students took from their education and experience at the colleges of the atom in the midst of a Cold War was the knowledge and assurance that they could administer their modernist state. Those engineers learned to run the technology of the atomic age via "containment, reactor instrumentation and control, and safety systems."[1] Students learned how to run the complex economic, social, and military infrastructure of the state that always threatened to run out of control. Looking back on the first twenty years of the reactor, then chairman of Michigan's nuclear engineering department Henry Gomberg celebrated the "distinguished body of scientists and engineers" the facility had trained. In contrast, the supervising health physicist Ardath Emmons chose to celebrate the "weekly meetings" which doled out everyone's "specific tasks to perform" and the excitement of erasing from the chalkboard "each item with its completion." Likewise, a Michigan graduate from the reactor training program and later a development manager for the construction giant Bechtel, Jerome Shapiro recalled mostly how "the control room seemed to be the center of attention" wherein "the regular cadence of registers [and] the painfully slow movement of the rods" occupied the students' long nighttime hours.[2] To the nation's scientific leadership, the reactor served to convey knowledge about the natural world and its usefulness; to local scientific observers, the reactor embodied a society in a Cold War, where order, tasks, and regularity were the sure paths for the United States to follow. Famously, the founder of the American nuclear navy Hyman Rickover insisted on highly trained crews to operate the reactors of his fleet. Elsewhere, the rapid expansion of the whole nuclear industry created a severe shortfall in expertise of operators, technicians, and engineers, and consequently nuclear power stations drew people from coal and oil plants with minimal training, creating the impression that nuclear power plants were "designed by geniuses and operated by idiots."[3] Unfair or not, such quips reveal that the creation of a technology, particularly one at the scale of a national industry, was really a problem of gaining widespread training and expertise.

Between 1949 and the late 1970s, nearly 150 university programs, accompanied by over two dozen new reactors dedicated to education and training, appeared on university campuses across the United States and around the world, to supply the experts for the new atomic age. Unlike their industrial or military counterparts, research and training reactors within the nation's colleges were generally small. When North Carolina State College in Raleigh built and ran the first university-based reactor, it was a light water reactor of only 10kw power. The University of Michigan's was a more substantial reactor for research and training, the 1000kw Ford Nuclear Reactor, opened in 1957. As noted above, the Ford Reactor was attached to the Michigan Memorial Phoenix Project. Historian Joe Martin noted that the Phoenix Project was an important early university-based institution for nuclear research, sponsoring seven new laboratories, more than 100 research projects, and supporting 20 yearly positions for faculty research and 60 graduate students over the following decade.[4] Among its achievements, the Phoenix Project funded the early development of the bubble chamber by Donald Glaser, for which he received the Nobel Prize in 1960, as well as H.R. Crane's study of carbon-14 dating.[5] As the founders of the Phoenix Project knew full well, nuclear reactors were the predicament of the postwar United States in a nutshell: new technology built by American industry from designs and expertise gained in public universities to reshape the domestic and international fabric of the nation itself. However, due to legal stipulations, no institution outside of the Atomic Energy Commission (AEC) could possess nuclear material in the United States. Thus, the specific desire for government control and secrecy over the atom faced off against the general political conservativism of American institutions to limit such actions. Consequently, the AEC resisted the expansion of their in-house training programs, favoring instead universities throughout the United States that developed local nuclear engineering training programs. Universities applied for and received equipment from the AEC, notably reactor fuel rods, mostly as a free loan. This generosity was a political middle ground between accommodating and resisting government control over atomic energy. Remarkably, it even extended to the numerous atomic programs the United States helped establish around the world: "The U.S. Atomic Energy Commission provides facilities and funds," stipulated AEC administrator Jesse Perkinson to Puerto Rico when establishing their Nuclear Center in 1958, for example.[6] What the AEC did not provide was expertise; hence the overt purpose of new nuclear centers in places like Puerto Rico, North Carolina State College, Iowa, MIT, Pennsylvania State College, and the University of Michigan was explicitly to train the next generation of nuclear engineers, specialists, and operators.

This chapter foregrounds those nuclear reactors designed, built, and maintained specifically for teaching. It argues that research and training reactors shaped both the technology of and professional attitudes toward nuclear power. Like historian Mark Bowles, I was amazed how little had been written about research and training reactors, not least because they were most often an undergraduate or graduate student's first point of contact with nuclear technology which led to lifetimes spent in the industry.[7]

Additionally, it was these facilities that most openly displayed the reactor glowing blue to school and scout groups as well as to the general public. Moreover, the skills and values of the era were forged in training reactors: no matter whether it was to support security through weapons development or search for cures through biomedical science, the era required a vast increase in trained nuclear scientists and engineers. "We cannot maintain an adequate level of sound scientific research without a strong body of highly trained and creative scientists," technocrat Lloyd Berkner said.[8] It was in the training reactors at places like the University of Michigan and North Carolina State College, and later at MIT, Missouri, Caltech, Stanford, Iowa, Georgia, and even Reed College (the only solely undergraduate-run reactor in the nation) that students took courses on nuclear science and engineering, did experiments for theses with nuclear reactors, and took courses to become the operators of their university's atomic facility.

The emergence of teaching reactors within the nation's colleges and universities represents one of the key legacies of the Manhattan Project. The Cold War period was hardly the first time education had shaped technology, and technology reshaped education. Broadly, beginning in the industrial revolution and continuing through the atomic age a new "technical literary" emerged.[9] To make sense of the world, both the traditional skills of reading and writing but also a new mathematical comprehension and scientific literary were demanded. To be a citizen of the Jeffersonian republic of the 18th century required land and letters; to be a citizen of the modern industrial world required mathematics; but to be a citizen of a superpower in the latter half of the twentieth science required science and engineering. Throughout, the training of a new scientific or engineering disciple became paramount.[10] The use of a teaching reactor to train the future generations of nuclear engineers replicated the pedagogy of mathematical physicists. Working through a world of pen and paper, historian Andrew Warwick not only charted the twin familiar loci of institution and discipline, Cambridge and Mathematical Physics, but also explained pedagogy as culture where scientific communities work through the problems of normal science by training new disciples.[11] Looking at labor, especially the labor for and of students, is critical to understanding the culture that binds technical communities together. The expansion of nuclear engineering as a subject and teaching reactors as a pedagogical tool helped create much of that culture. As AEC commissioner Wilard Libby said to the American Nuclear Society of NYU in 1956, "The world will more and more belong to the scientifically educated man—the engineer, the physicist, the chemist, the biologists—who, through training and knowledge can create and master the new technologies which increasingly are forming the broad base on which our civilization rests."[12] The emergence of the teaching reactor, then, had clear resonances with the far-reaching pedagogical changes in Cold War science. The rise of Feynman diagrams, David Kaiser argued, became a particular pedagogical strategy to produce mechanical physicists in the face of a veritable flood of students, where too many new faces in too many new departments broke down the traditional and much romanticized ways of personal mentorship.[13] More than any other, the act of apprenticeship in graduate schools significantly imparts the social structure, mores, and ideals of every scientific and technical community; training, as Michel Foucault noted, is both about education and control.[14] For the nuclear engineers, the incorporation of nuclear engineering within established modes of engineering shaped what a nuclear reactor could be and how it could work. At the same time, the training site of the swimming pool teaching reactor shaped what a nuclear engineer could be and how he could work.

THE POWER TO BUILD THE WORLD

The atomic bomb burst into worldwide consciousness in August 1945. Beyond the Manhattan Project and the deeply held myth that the atomic bomb ended World War II, what the project produced was, first and foremost, thousands of atomic and hydrogen bombs; hundreds of missile silos, submarines, and bases; and dozens of vehicles and systems to ensure nuclear weapons were deployed.[15] The dramatic nature of atomic tests, the spectacle of mushroom clouds, and the political frenzy of missile and bomber gaps always threatened to turn the Cold War with the Soviet Union hot.[16] Optimistically, the postwar civilian Atomic Energy Commission ran a radioisotope distribution project to put a human, and humane, face on the birth of the atomic age with the hype of atomic bomb swords turned into medical plowshares.[17] Equally optimistically, nuclear power reactors spread across the country on the promise of electricity 'too cheap to meter.' When Eisenhower launched Atoms for Peace in 1953, there were 30 operating reactors in the United States, and rather less than that number in the rest of the world. Five decades later, as George W. Bush struggled to defend a superpower against terrorists, 439 reactors were operating across 31 countries.[18] As historian Sean Johnston noted, while myriad tomes detail the experiences of the atomic scientists, the role of nuclear engineers in the development of the atomic bomb, the first atomic piles, and the vast enterprise of the nuclear industry remain understudied and underappreciated.[19] Indeed, spanning half a century, that is a story of American power, in both senses, during the Cold War. As historian David Nye argued, "America's energy abundance was understood to be the result of capitalism, and the scarcity of cars and appliances in the communist countries was seen as indicative. The famous kitchen debate between Richard Nixon and Nikita Khrushchev over the relative merits of capitalism and communism could not have been held in a more appropriate location."[20] It is important to keep in mind, however, that such efforts paled before the production of weapons: Gordon Dean, 1950-53 chairman of the AEC, mentioned that the atomic energy project cost about ten times the price of the Panama Canal, some $9 billion compared to $366 million, of which the vast majority went to weapons plant infrastructure, including infrastructure like the new Portsmouth plant with a floor area greater than Ford's Willow Run assembly plant and the Pentagon combined, that consumed more electricity than New York City, and utilized a construction workforce of 65,000 people or "about five percent of the total construction force of the nation."[21]

Undergirding the multiple infrastructures of weapons, power plants, and radioisotopes were legions of trained nuclear engineers, scientists, technologists, and technicians. Historian John Krige persuasively argued that the education of desire for the peaceful atom was part of the overarching Cold War era context of the entire Atoms-for-Peace movement. Years of secrecy had held back American efforts to share nuclear technology, but the accelerating process of decolonization, accompanied by the realignment of new states toward either the Soviet or American spheres of influence gave a new urgency to international cooperation.[22] Moving beyond piecemeal experience, Oak Ridge hosted both the Reactor Training School and the Radioisotope Training Program, which by 1953 had already coached 700 people from over 400 institutions and 30 nations in the uses of radioisotopes.[23] Though radioisotopes flowed liberally outwards and bombs were stockpiled, the AEC resisted further expansion of its in-house training regimes for fear of not being appropriately conservative enough. Though the AEC continued to have many in-house training courses and was connected to regional institutes that provided education under contract, "the prime responsibility of training these scientists and engineers…is the province of the universities and colleges."[24] As anxious as the AEC was to try and maintain the separation between State and Atom, when a group of university nuclear engineers met in 1959 to discuss the problems of establishing teaching and research reactors on college campuses, their chair specifically noted that theirs was "a strictly non-AEC group [s]o these points can come up freely."[25]

The Cold War era witnessed institutions of higher learning reconfigured to train the future generations of scientists and engineers it was assumed the state would need in perpetuity. Universities eagerly embraced their educational task as a patriotic and profitable mission. They expanded via large research contracts but also in response to the demands to recruit and train the talent the Cold War required. In 1956, Willard Libby spoke before the second student chapter of the American Nuclear Society at New York University and said, "to my mind, the most important bottleneck in the peaceful application of atomic energy is not money…it is *trained manpower*." Little seemed more promising for young people with scientific and engineering skills than the expansive horizon of atomic energy, Libby noted. Gesturing to his audience, he said that theirs was "an era of rapid development in the peaceful uses of atomic energy in which our annual needs for technical personnel will mount steadily higher."[26] While the scientists and engineers created bombs and reactors, their teachers created syllabi, textbooks, exams, and practicums that, in deep ways, shaped the ways everyone approached the era. Most significantly, in setting up training programs in universities, nuclear engineering leaders insisted that they have a reactor. "In the whole field of nuclear technology," John Trimmer advised universities, "the reactor is central, the key device, the natural focus around which subsidiary instruments, great or small…, may best be grouped."[27]

Pedagogically, the focus of training was the reactor. "Teaching of nuclear technology can well begin with the study of the reactor itself. It has great intrinsic interest, combining simplicity of basic concepts with complex variety of eventual design possibilities." For any technologically advanced society, where scientists and engineers would remain in high demand and where complex systems would be everywhere, training via a reactor would be doubly beneficial. "As a physical system," Trimmer said, "the reactor is an intriguing example for courses in system design, instrumentation, and automatic control."[28] In short, the reactor not only powered society, it was to be its essential organizing shape. The paradigm of modern technocratic society with its systems of control, feedback, and automation would be established through a regime of training in nuclear reactors.

The case of the University of Michigan, for example, provides a nice illustration. The institution asked the AEC for a low-cost, 1000kw, swimming pool reactor, with a peak neutron flux of 10^{13} n/cm^2/sec. With an estimated cost of $800,000, the design was essentially a modified copy of Oak Ridge's Low Intensity Training Reactor. Michigan had decided against using either a copy of the 'North American Reactor,' because it promised only 200kw and just a 10^{12} neutron flux density, or replicating the 'water boiler' reactor already being installed at the North Carolina State College. Michigan questioned the water boiler's promise of a high neutron flux rate at seemingly amazingly low power (10^{13} n at only 30kw) and opted instead, they said, for a "larger, somewhat more costly system whose performance can be assured."[29] Undoubtedly a product of the excitement of nuclear science and engineering as a field, the local philanthropic culture of Michigan, and the desire of major corporations to get into atomic work, the Ford Motor Company donated $1 million to build the University of Michigan's nuclear reactor in May 1953.

THE NORMAL HALF LIFE OF A GRADUATE STUDENT

The new cadres of nuclear engineers and their teachers understood that nuclear reactors were not merely representative of the Cold War era, they essentially were the Cold War. Historian Helge Kragh succinctly summed up that, if physicists had created the atomic bomb, it was the engineers who created the atomic era: "By 1958, with the publication of Alvin Weinberg's and Eugene Wigner's authoritative *The Physical Theory of Neutron Chain Reactors*, the work of the physicists was over in this area. A new species of engineer-scientist, nuclear engineers, were responsible for the second phase of the development of nuclear reactors."[30] The creation of hundreds of new departments of nuclear engineering, new societies of nuclear engineering, new journals, and new teaching reactors meant that the incorporation of the atomic into engineering represented one of the greatest changes to the profession in a century. Advising young engineers on their professional identity and responsibilities, John Constance said that "the Atomic Age is making, perhaps, its greatest impact on engineering talents and resources."[31] It was little exaggeration that, in only 20 years, the idea and profession of engineers of all stripes had been radically altered by the demands of the atomic age to create its reactors, plants, processes, and bombs. To historian Sean Johnston, the core challenge of the atomic era for engineers "were shifting conceptions of the terms 'nuclear' and 'engineer'," and to grasp the Cold War we need to explain how those "understandings were constructed in distinct places and times."[32] One historical problem is that the era's scientists and engineers did not see resonance of their technical endeavor with the larger social context which supported it and informed it. Through seminal studies like Crosbie Smith and M. Norton Wise, *Energy and Empire: A Biographical Study of Lord Kelvin*, it is clear to historians that scientific knowledge is shaped through technological concerns.[33]

In the United States, almost all nuclear reactors for civilian power were light-water reactors. Even by the 1950s, the principles of building and running a light-water reactor were known and stable. They also comprised most of the 'test reactors' as well, which usually offered extremely high neutron fluxes for studies with very high radiation levels, most famously the Plum Brook test reactor in Ohio, which did the early design experiments for NASA's nuclear rocket engine.[34] As James Mahaffey points out, the designs for light-water reactors stemmed from the remarkable work of Hyman Rickover to develop a safe, compact, and reliable nuclear engine for submarines.[35] Ordinary water both cools and moderates the reaction, and while the core could melt down and there was the serious risk of hydrogen and oxygen bubbles forming that might explode, the light-water reactor had no graphite or sodium that would create radioactive clouds as it exploded. To get more power out of the system, however, and to limit its space requirement, the water was pressurized, which meant that the boiling point of the water was increased.

At least twenty reactors of various designs eventually existed across the United States, from the Low Power Water Boiler (LOPO) light-water moderated reactor at Los Alamos, operating at 1/20th of a watt, through the Experimental Breeder Reactor liquid metal-cooled reactor at the Reactor Testing Station in Idaho operating at 100kw, to the graphite moderated, water-cooled, Hanford Reactor in Washington State operating at 1000MW, producing plutonium.[36] The reactor on which a vast amount of hope and effort was placed was the "breeder reactor," a reactor that would create more power and material than it used. If the process could be mastered, an initial unit of uranium would be alchemically transformed into plutonium generating heat for power, and as uranium was fed in, plutonium could be extracted and used once again to produce even more heat for power. To the young nuclear engineer of the 1950s, however, these power reactors lay in the distant future.

A student's first and probably most influential experience was with another type of nuclear reactor, the reactor for education and training. Normally space was not important, especially for larger state universities, and no pressure housing was used because students needed to access the teaching reactors. With few boundaries, universities invested in the "swimming pool" reactor. In a swimming pool reactor the fissioning core is completely surrounded by water, meaning that scientists and students could stand above the pool and gaze upon the blue glow, the only visible sign that nuclear fission was going on before their eyes. To occasional visitors, the blue glow was a "wonderment" but to the properly educated and trained scientists and engineers of the atomic age, the reactor was "unremarkable" as it was simply a "tool."[37] The blue glow of the reactor, visible only in the swimming pool design, stemmed from Cerenkov radiation, which emanates from nuclear cores encased in water because the velocity of some electrons is greater than the velocity of light through the water. This visibility of the reactor core was important in the 1950s and 1960s, as expansion of training regimes went forward to teach the legions of engineers, technicians, and maintenance workers the new nuclear future would require.

Critically, "the reactor design" of the University of Michigan's nuclear reactor, per Henry Gomberg, "had to fulfill the criteria for a good teaching device. [One] point we felt was quite important: visual access, the ability of the student to observe an operation as much of the time as possible."[38] Students could introduce medical or material samples into the core to be irradiated or tested, while staff could lower control rods into the core to slow or halt the nuclear reaction as the control rods absorbed the neutrons needed to sustain a nuclear chain reaction. In addition, a number of nuclear engineering educators reveled in the opportunity that "students can play with and take apart and put back together [a swimming pool reactor] without really having any hazardous operation," though that very flexibility placed "a real severe limitation on what a real nuclear reactor can do."[39]

The Michigan College of Engineering's nuclear engineering program was built around this understanding of pedagogical instrument and purpose. Michigan lauded its novel educational offering. While students in the program took "courses [that] deal primarily with theoretical aspects of nuclear science," replicating the emergent programs at North Carolina State and MIT, the opening of the Ford reactor permitted the expansion of practical training in reactor design and operation. As the manual proclaimed, courses "give the student an opportunity to obtain a practical acquaintance with reactors." That practical acquaintance took the shape of training students to utilize the reactor's radiation for experimental work in radiation spectra, neutron scattering and absorption, flux distribution, neutron shielding, and reactor kinetics. "In addition," the manual continued, "the reactor will be used for instruction in connection with other courses (e.g., chemistry, physics, engineering, medicine, and, perhaps, some of the natural sciences) in which a basic knowledge of reactor technology is desirable." The training above all concentrated on "familiarizing students with reactor technology."[40]

In this way, Michigan committed itself to "develop the soundest possible training program for engineers and scientists in this new, rapidly expanding field. To attract the best men, to stimulate their interest, and to provide the proper training, the basic tool for such a program must be available."[41] In other words, the reactor housed not only an active fission reaction in a swimming pool but also the ability to attract and define the "best" men of the age, university educated nuclear engineers, as well as establish the emergent meaning of what was considered "proper" training as engineering looked to the future of nuclear energy control and utilization. The reactor itself was not a power reactor but a pedagogical reactor, instilling in its graduate students an epistemology of the atomic; the structure of how to deal with new technologies, new forces of energy, and the new social shape of the atomic age. The "best men" had to be trained to take "control" of the technology of the reactor and the products of the reaction. As engineers they would extend a theoretical understanding of nuclear energy into a practical employment of the "basic tool," the reactor itself, to solve future "new needs and problems."[42]

The teaching reactor explicitly shaped the student's expectations about the nature of those new problems. The reactor itself was designed to accommodate the schedule of a university education, specifically a graduate education. In what Henry Gomberg called 'the normal half-life of a graduate student,' he reiterated to his colleague at the Penn State Reactor Conference how and why a training reactor functioned:

> This is a story we have told many times … If you are going to do solid state work, genetic work, and other work of this sort, and you want to do it with graduate students, you have to give them an opportunity to get their data in a reasonable length of time. With a flux of about 10^{13} you can reach 10^{20} integrated flux, which is more or less accepted in the field as a good test value, in a reasonable time. This gives the student the opportunity of designing his experiment, building his equipment, gathering data and analyzing it in what we like to call the normal half-life of a graduate student. … we are talking about 10^6 seconds. … If it is a week or ten days, this is a completely reasonable time for a student to run an experiment. This is one of the principal reasons that we insist on this flux level.[43]

THE EDUCATED NUCLEAR ENGINEER

The nuclear engineer walked proudly in the Cold War era. For many he was an extension of traditional engineering, which in itself implied that the nuclear age was not a fundamental break. For others, he was an amalgam of expertise, much like the reactor itself, made up in varied parts from mechanical engineers building the reactor housing, thermal engineers building heat transfer systems, chemical engineers learning about the properties of uranium, thorium, and plutonium as fuel and about the bizarre and unexpected chemical properties of the numerous by-products of fissions. In what would become the most infamous part of the nuclear industry, "control engineers," historian Sean Johnston explains, "had the novel task of designing fail-safe systems to ensure stability of the finely balanced chain reaction—a responsibility more difficult than initially anticipated."[44] At the same time, the nuclear engineer was a political and social amalgam: heir to American's heroic past of visionary engineers and father to America's future in the atomic age. Revealing the struggles of others to gain entry into engineering has been the task of Amy Slaton's important work on African American engineers and Amy Bix's significant history of women entering engineering.[45] At least as important as race and gender, class shaped the nuclear engineers' lives. When AEC Chairman David Lilienthal wanted to remove the security fences around the civilian areas at Oak Ridge, he unexpectedly encountered a local protest to keep the fences: "these citizens had no desire to surrender what they deemed a mark of superiority, something that set them off as a special, separate, a privileged caste."[46]

The nuclear engineers possessed both the technical details of mastering the atom's power and the cultural codes of entry into a modern, elite community. In an optimistic vision, the postwar president of the University of Michigan, Harlon Hatcher, said, "it is the obligation of education to move with all energy and speed to dissipate the dark clouds of gloom and terror that surround nuclear fission as a physical fact, to train and direct men in the vision and the skill to move its power into peaceful and beneficent applications, and to adjust our social and governmental agencies to absorb the new dimensions of life."[47] Iconic of a revolutionary new innovative technology, and seemingly destined to spur myriad fantastical applications and gadgets, nuclear power was ultimately best summed up by Hyman Rickover that while "too much emphasis" had been placed on innovation, "not nearly enough on the daily drudgery of seeing that every aspect…is in fact properly handled every day."[48] As historians, sociologists, and anthropologists of the modern technological world have recognized, the maintainers do all the unappreciated work of keeping society running while the innovators get all the glory.[49] To David Lilienthal, writing the year after Three Mile Island, it was the nuclear engineers who "gingerly executed a series of operations to cool down the plant's dangerously overheated nuclear core."[50] The engineers did not cause the crisis, they instead possessed the skill to moderate the crisis; the cool and unflappable engineer as hero. Yet, the overwhelming popular images of deviant and rebellious youth culture of the 1950s contrasts with the serious young men who became nuclear engineers. As Margot Henriksen described in a host of novels, films, and magazines, American society confronted a conflict between older and younger citizens. It was a battle over dress, modes of speech and address, and the nuclearity of the family and the home: when in the movie *Rebel Without a Cause,* James Dean's character screams at his parents, "*You're tearing me apart,*" the parental understanding of such outbursts is that it is a product of the age, "The Atomic Age."[51] But it is also too easy to cast a youth wearing a leather jacket against others wearing white shirts and pocket protectors. Allen Ginsberg screamed that America should "Go fuck yourself with your atomic bomb," but (in decidedly less inflammatory language) so really did the serious new nuclear engineers. The nuclear engineers at Michigan or North Carolina State or, notably, Reed College, did not automatically condemn atomic power alongside the atomic bomb. Those students may have more accepted Nevil Shute's remedy to the atomic age over Ginsburg's: toward the end of Shute's famous apocalyptic 1957 novel *On the Beach*, as a husband and wife witness the end of life on Earth and ask how it could have happened, the husband speculates that "the only possible hope would have been to educate them out of their silliness."[52]

Notes

1. Richard Rabbideau, "University Reactor Set Off Yesterday." *The Michigan Daily*, September 16, 1957, 10. Helen J. Lum, "Twentieth Anniversary of the Ford Nuclear Reactor," October 1977. In folder "Phoenix. History, 20th anniversary," box 5. Michigan Memorial Phoenix Project papers, Bentley Historical Library, University of Michigan.

2. Untitled notes and quotes. n.d. In folder "Phoenix. History, 20th anniversary," box 5. Michigan Memorial Phoenix Project papers, Bentley Historical Library, University of Michigan.

3. Richard Rhodes, *Nuclear Renewal: Common Sense About Energy* (New York: Whittle Books, 1993), 49.

4. Joseph D. Martin, "The Peaceful Atom Comes to Campus," *Physics Today* 69:2 (2016): 40–46.

5. Jens Zorn, ed., *On the History of Physics at Michigan* (Ann Arbor, MI: University of Michigan, 1988), 51. Peter Galison, "Bubble Chambers and the Experimental Workplace," in *Observation, Experiment, and Hypothesis in Modern Physical Science,* ed. Peter Achinstein and Owen Hannaway (Cambridge, UK: Cambridge University Press, 1985), 309–373.

6. Summary of talk by Jesse Perkinson of the AEC by G. Hoyt Whipple, "Report on Visit to the Commonwealth of Puerto Rico," June 1958. In Folder "Puerto Rico: Report." Box 3. Henry J. Gomberg papers, Bentley Historical Library, University of Michigan, 5.

7. Mark D. Bowles, *Science in Flux: NASA's Nuclear Program at Plum Brook, 1955–2005* (Washington, D.C.: NASA, 2006), xxv.

8. L. V. Berkner, *The Scientific Age: The Impact of Science on Society* (Yale University Press, 1964), 16.

9. Edward W. Stevens, Jr., *The Grammar of the Machine: Technical Literary and Early Industrial Expansion in the United States* (New Haven, CT: Yale University Press, 1995), 2.

10. Robert Kohler places students equally alongside "authority and access to tools and craft knowledge." Kohler, *Lords of the Fly*, 12; David Kaiser, "Moving pedagogy from the periphery to the center," in David Kaiser ed. *Pedagogy and the Practice of Science: Historical and Contemporary Perspectives* (Cambridge, UK: Cambridge University Press, 2005), 1–8.

11. Andrew Warwick, *Masters of Theory: Cambridge and the Rise of Mathematical Physics* (Chicago: University of Chicago Press, 2003). Also, *Pedagogy and the Practice of Science: Historical and Contemporary Perspectives,* ed. David Kaiser and Andrew Warwick (Cambridge, UK: Cambridge University Press, 2005); Catherine Jackson, "Visible Work: The Role of Students in the Creation of Liebig's Giesson Research School," *Notes & Records of the Royal Society* 62 (2008): 31–49; Kathryn M. Olesko, *Physics as a Calling: Discipline and Practice in the Königsberg Seminar for Physics* (Ithaca, NY: Cornell University Press, 1991).

12. AEC press release: "Remarks by Willard F. Libby to the American Nuclear Society, NYU." April 27, 1956. In Box 593, David E. Lilienthal papers, Princeton University Library.

13. David Kaiser, "Cold War requisitions, scientific manpower, and the production of American physicists after World War II," *Historical Studies in the Physical Sciences*, 33:1 (2002), 131–159; Also Stuart W. Leslie, *Cold War and American Science: The Military-industrial-academic Complex at MIT and Stanford* (New York: Columbia University Press, 1993), 10.

14. "The examination is the technique by which power, instead of emitting the signs of its potency, instead of imposing a mark on its subjects, holds them in a mechanism of objectification." Michel Foucault, *Discipline and Punish: The Birth of the Prison* (New York: Penguin, 1991), 187.

15. Richard Rhodes, *The Making of the Atomic Bomb* (Penguin Books, 1988); Eric Schlosser, *Command and Control: Nuclear Weapons, the Damascus Accident, and the Illusion of Safety* (New York: Penguin, 2014). For the development of the ICBM, see Thomas P. Hughes, *Rescuing Prometheus: Four Monumental Projects that Changed the Modern World* (New York: Vintage, 2000), chapter 3. Hughes later noted that even such figures in the order of hundreds of millions of 1960s dollars are still an order of magnitude below the defense contracts given to aerospace industries, Lockheed alone being awarded over $10 *billion* between 1960 and 1967. Thomas P. Hughes, *Human-Built World: How to Think about Technology and Culture* (Chicago: University of Chicago Press, 2005), 81. Nor has the end of the Cold War halted such expensive development of dubious scientific projects for the military. The $12 million-per-year research budget on the Hafnium grenade, a "miniature [atomic] bomb. Explosive yield, 2 kilotons. Size, five-inch diameter," for example, is ongoing. See Sharon Weinberger, *Imaginary Weapons: A Journey through the Pentagon's Scientific Underworld* (New York: Nation Books, 2006), 10. Also ongoing is the creation of military superheroes by the MIT Institute for Soldier Nanotechnologies, which was "kick-started" by a comparatively paltry $50 million from the U.S. Army Research Office in 2002. See Colin Milburn, "Nanowarriors: Soldier Nanotechnology and Comic Books," *Intertexts* 9 (2005): 77–103, 77. The flipside of nuclear weapons was the equal insanity of civil defense, see Scott Gabriel Knowles, *The Disaster Experts: Mastering Risks in Modern America* (Philadelphia: University of Pennsylvania Press, 2012), chapter 4.

16. Robert Jay Lifton and Greg Mitchell, *Hiroshima in America: Fifty Years of Denial* (New York: Grosset/Putnam, 1995); Gregg Herken, *The Winning Weapon: The Atomic Bomb in the Cold War, 1945–1950* (New York: Alfred A. Knopf, 1980).

17. Angela N. H. Creager, *Life Atomic: A History of Radioisotopes in Science and Medicine* (Chicago: University of Chicago Press, 2013), 154.

18. Constance Perin, *Shouldering Risks: The Culture of Control in the Nuclear Power Industry* (Princeton, NJ: Princeton University Press, 2005), xvii–xviii.

19. Sean Johnston, *The Neutron's Children: Nuclear Engineers and the Shaping of Identity* (Oxford: Oxford University Press, 2012), 7.

20. David E. Nye, *Consuming Power: A Social History of American Energies* (Cambridge, MA: MIT Press, 1998), 202.

21. Gordon E. Dean, *Report on the Atom: What You Should Know about the Atomic Energy Program of the United States* (New York: Alfred A. Knopf, 1954), 68–69.

22. John Krige, "Techno-Utopian Dreams, Techno-Political Realities: The Education of Desire for the Peaceful Atom," in *Utopia-Dystopia: Conditions of Historical Possibility,* ed. Michael D. Gordin, Helen Tilley, and Gyan Prakash (Princeton, NJ: Princeton University Press, 2010): 151–175.

23. Gordon E. Dean, *Report on the Atom: What You Should Know about the Atomic Energy Program of the United States* (New York: Alfred A. Knopf, 1954), 220–21.

24. AEC press release: "Remarks by Willard F. Libby to the American Nuclear Society, NYU." April 27, 1956. In Box 593, David E. Lilienthal papers, Princeton University Library.

25. Transcript of "Proceedings of the Meeting of the Members of and Consultants to the Subcommittee on Research Reactors," March 28 and 29, 1959. Folder "Proceedings of the Meeting of the Subcommittee on Research Reactors, March 28 and 29, 1959," Box 12. Michigan Memorial Phoenix Project papers, Bentley Historical Library, University of Michigan, 17.

26. AEC press release: "Remarks by Willard F. Libby to the American Nuclear Society, NYU." April 27, 1956. In Box 593, David E. Lilienthal papers, Princeton University Library.

27. John Trimmer, "Availability and Selection," *Nuclear Reactors for Industry and Universities,* ed. Ernest Henry Wakefield and Clifford K. Beck (Pittsburgh: Instruments Publishing Company, 1954), 11.

28. Ibid.

29. "Proposal for a Nuclear Reactor at the University of Michigan," January 22, 1952. Folder "Phoenix. Topical 2. AEC. Ford Nuclear Reactor," box 12, Michigan Memorial Phoenix Project papers, Bentley Historical Library, University of Michigan, 5.

30. Helge Kragh, *Quantum Generations: A History of Physics in the Twentieth Century* (Princeton, NJ: Princeton University Press, 1999), 285.

31. John D. Constance, *How to Become a Professional Engineer* (New York: McGraw-Hill, 1966), 51.

32. Johnston, *The Neutron's Children*, 14.

33. Crosbie Smith and M. Norton Wise, *Energy and Empire: A Biographical Study of Lord Kelvin* (Cambridge, UK: Cambridge University Press, 2009). Also Paul Forman, "The Primacy of Science in Modernity, of Technology in Postmodernity and of Ideology in the History of Technology," *History and Technology* 23 (2007): 1–152.

34. Mark D. Bowles, *Science in Flux: NASA's Nuclear Program at Plum Brook, 1955–2005* (Washington, D.C.: NASA, 2006).

35. James Mahaffey, *Atomic Awakening: A New Look at the History and Future of Nuclear Power* (New York: Pegasus Books, 2009), 188–89, 213–21.

36. Taken from William Breazeale, "Nuclear Reactor Types," *Nuclear Reactors for Industry and Universities,* ed. Ernest Henry Wakefield and Clifford K. Beck (Pittsburgh: Instruments Publishing Company, 1954), 6–8.

37. Helen J. Lum, "Twentieth Anniversary of the Ford Nuclear Reactor," October 1977. In folder "Phoenix. History, 20th anniversary," box 5. Michigan Memorial Phoenix Project papers, Bentley Historical Library, University of Michigan.

38. Transcript of "Proceedings of the Meeting of the Members of and Consultants to the Subcommittee on Research Reactors," March 28 and 29, 1959. Folder "Proceedings of the Meeting of the Subcommittee on Research Reactors, March 28 and 29 1959," Box 12. Michigan Memorial Phoenix Project papers, Bentley Historical Library, University of Michigan, 9.

39. Transcript of "Proceedings of the Meeting of the Members of and Consultants to the Subcommittee on Research Reactors," March 28 and 29, 1959. Folder "Proceedings of the Meeting of the Subcommittee on Research Reactors, March 28 and 29, 1959," Box 12. Michigan Memorial Phoenix Project papers, Bentley Historical Library, University of Michigan, 5.

40. *The Ford Nuclear Reactor: Description and Operation.* June 1957. Folder "The Ford Nuclear Reactor: Description and Operation, 1957," Box 24, University of Michigan College of Engineering papers, Bentley Historical Library, University of Michigan, II-1.

41. *The Ford Nuclear Reactor: Description and Operation.* June 1957. Folder "The Ford Nuclear Reactor: Description and Operation, 1957," Box 24, University of Michigan College of Engineering papers, Bentley Historical Library, University of Michigan, II-2.

42. *The Ford Nuclear Reactor: Description and Operation.* June 1957. Folder "The Ford Nuclear Reactor: Description and Operation, 1957," Box 24, University of Michigan College of Engineering papers, Bentley Historical Library, University of Michigan, II-1.

43. Comments by Dr. Gomberg as part of the transcript of the Penn State Reactor Conference. Attached to Letter from Lyle Borst to H. J. Gomberg, October 10, 1955. Folder "Phoenix. Topical 2. Committees. Subcomm. on Unclassified Nuclear Reactors," box 12, Michigan Memorial Phoenix Project papers, Bentley Historical Library, University of Michigan, 134–35.

44. Johnston, *The Neutron's Children,* 5.

45. Amy E. Slaton, *Race, Rigor, and Selectivity in U.S. Engineering: The History of an Occupational Color Line* (Cambridge, MA: Harvard University Press, 2010), and Amy Sue Bix, *Girls Coming to Tech!: The History of American Engineering Education for Women* (Cambridge, MA: MIT Press, 2013). Also Tracy Kidder, *The Soul of a New Machine* (Boston: Little, Brown, and Co.: 1981)

46. David E. Lilienthal, *Atomic Energy: A New Start* (New York: Harper and Row, 1980), 72.

47. Harlon Hatcher, "The Impact of Nuclear Energy on Education." Speech at a conference "The Social Impact of Nuclear Energy," June 24, 1954. In Folder "Phoenix. Topical 1. The Atom Reports, 1953–54," box 8, Michigan Memorial Phoenix Project papers, Bentley Historical Library, University of Michigan.

48. Constance Perin, *Shouldering Risks: The Culture of Control in the Nuclear Power Industry* (Princeton, NJ: Princeton University Press, 2005), 26.

49. David Edgerton, *The Shock of the Old: Technology and Global History Since 1900* (Oxford: Oxford University Press, 2006); Andrew Russell and Lee Vinsel, "Hail the Maintainers." Aeon.com, April 7, 2016.

50. David E. Lilienthal, *Atomic Energy: A New Start* (New York: Harper & Row, 1980), xii.

51. Quoted in Margot A. Henriksen, *Dr. Strangelove's America: Society and Culture in the Atomic Age* (Berkeley: University of California Press, 1997), 163. See pp. 162–167 for the fuller analysis.

52. Nevil Shute, *On the Beach* (New York: Vintage, 2010), 301.

CHAPTER SIX

Political Scientists
The Atomic Scientists and the Emergence of a Politically Engaged Scientific Community

Ian Graig

O n Earth Day 2017, thousands of scientists marched in Washington and other cities across the United States to call on the Trump administration to support the sciences and adopt evidence-based policies on such issues as climate change. In his story about the March for Science, Nicholas St. Fleur of the *New York Times* wrote that the marchers were "abandoning a tradition of keeping the sciences out of politics."[1] But while the scientific community has long strived to be apolitical, the March for Science hardly marked the first instance of scientists engaging in political activity to influence public policy.

One of the clear precursors of the March for Science can be found in the political activism of the physicists, chemists, and other scientists who worked on the Manhattan Project. Some of the most prominent of those scientists first became involved in the policy process through their efforts early in World War II to convince British and American political leaders to launch the project at a time when the atomic bomb was little more than a theoretical concept. A larger group later played a leading role in trying to convince American and British political leaders to take steps to ensure the control of nuclear weapons. While they were hardly the first political scientists, the efforts of the Manhattan Project scientists influenced the emergence of a more politically engaged scientific community in the United States during the years after World War II.

The "atomic scientists" movement remains of interest today, even seven decades after it burst on the national political stage at the dawn of the nuclear age. The movement was the subject of several excellent studies by historians during the Cold War, particularly in the 1970s and 1980s when fear of nuclear war still ran strong and historians gained greater access to the documentary history of the Manhattan Project.[2] Memories of and interest in that movement faded in the post-Cold War era, however, as attention turned away from nuclear arms and scientific expertise became more accepted as a part of policy-making. For many, the atomic scientists' movement is now a distant memory, one marked by black-and-white photographs of earnest and generally bespectacled young physicists walking the corridors of the U.S. Capitol. But that movement remains one of the important legacies of the Manhattan Project, one whose renewed relevance has seemed especially clear since Donald Trump became president. The story of that movement is an essential part of this reflection on the first 75 years of the nuclear era.

This essay reviews the historical record of the scientists' ultimately unsuccessful efforts to get political leaders to embrace their ideas for international control of atomic weapons, and their more successful engagement in the debate over domestic atomic energy legislation. The story of the period during and immediately following the war, when the scientists' movement had its greatest influence, will serve as a starting point for looking at the movement's broader political legacy. Scientists played a relatively limited organized role in political discourse in the United States before World War II, most often when applying their technical expertise to discussions of narrow interest to their profession. The Manhattan Project scientists pushed beyond those boundaries, engaging policy makers at the highest levels of the U.S. government on issues that were fundamentally political in nature. Their efforts forced consideration of issues and ideas that might never have been debated, and forced policy makers who were not trained in the sciences to question, or at least defend, many of their basic assumptions. These efforts marked a watershed in the postwar political engagement of the scientific community. As the 2017 March for Science demonstrated, that engagement continues nearly 75 years after the Manhattan Project succeeded in developing the world's first atomic bomb.

Science: The Master of Man?

Historian Henry Adams, the grandson of John Quincy Adams, worried at the time of the American Civil War that science would someday give the human race the tools to "commit suicide by blowing up the world."[3] The first atomic bomb, which was detonated on July 16, 1945, near Alamogordo, New Mexico, unleashed more destructive force than Henry Adams could have possibly imagined. Within a month, two atomic bombs were used to destroy the Japanese cities of Hiroshima and Nagasaki. "With the discovery of atomic explosives," William H. McNeil wrote in his history of technology and warfare, "human destructive power reached a new, suicidal level, surpassing previous limits to all but an unimaginable degree."[4] Had science finally succeeded in giving mankind the power to commit suicide, as Henry Adams feared? Many of those who worked on the Manhattan Project were haunted by that question well before the Alamogordo test. An understanding of the physical forces at work in nuclear fission allowed them to glimpse, well before others, the potential power of a weapon that successfully harnessed the atom's vast stores of energy. That unique perspective drove many of those scientists to political activism, though it may also have led them to overestimate the bomb's ultimate *political* power.

While the scientists feared the implications of nuclear fission's military power, they also harbored hope that the energy released in fission could be harnessed for commercial, industrial, and even medical benefit. Many of the most politically active among them had an unshakeable faith in the ultimate good of science. That faith permeates the writings of Nobel Prize-winning physicist Niels Bohr, for example, who argued in a column published just five days after Hiroshima that a scientific achievement as remarkable as the discovery of fission must in the end yield positive benefits because "the progress of science and the advance of civilization have remained most intimately interwoven."[5] Bohr was not alone in his belief that the Manhattan Project, as the culmination of a period of unprecedented scientific discovery, could open the door to a whole new scientific era. The Manhattan Project itself was one of history's greatest gatherings of scientific minds, an enormous scientific and industrial enterprise that brought together the most remarkable group of scientists ever assembled for a common purpose, military or otherwise. The undertaking had a "whole magic mountain atmosphere," as physicist

Victor Weisskopf later recalled of Los Alamos, "of there being an elite society secluded from the rest."[6]

Yet neither faith in science nor belief in the importance of defeating Germany and Japan could overcome the fears of those who worked on the Manhattan Project that the development of atomic weapons posed a very real threat to mankind. Having come of age in a world torn apart by war, they understood the terrible human toll that could be extracted when science was used to develop new weapons. Many of those scientists consequently worked with equal diligence both to overcome the scientific and technological problems involved in developing atomic weapons and to address the political challenges, and potential opportunities, created by the development of such weapons.

THE WARTIME PUSH FOR AN OPEN WORLD

The scientists' efforts to address those political challenges first emerged in a wartime attempt to convince political leaders to take unprecedented steps to control the production and use of atomic weapons. Those efforts began in 1943 and 1944, when Niels Bohr approached American and British political leaders in an effort to convince them that the weapon being developed by the Manhattan Project would be of such destructive power that it should be taken out of the hands of national political leaders and placed under the control of an international organization. That organization could then also fully develop atomic energy for peaceful purposes. Bohr was a legendary figure in the world of physics and a relentless optimist. He came to the United States during the war largely to argue for international control, later explaining that "they didn't need my help in making the atom bomb."[7] He approached Allied leaders to raise his concerns well before the war's end because he feared that a contentious rivalry between the United States and the Soviet Union was likely to emerge as soon as the Allies defeated their wartime adversaries. Bohr presciently feared a dangerous postwar atomic arms race between the world's two great ideological adversaries. He assumed that Moscow would quickly develop its own atomic weapons because the fundamental concepts of nuclear fission were well known to scientists in the Soviet Union, which could also employ the industrial might needed to build a bomb within a few years.

Bohr thus began to work nearly eighteen months *before* the Alamogordo test to gain support from allied political leaders for an international agreement to control atomic weapons. Through lengthy memorandums and face-to-face meetings he attempted to persuade Franklin Roosevelt and Winston Churchill to support his proposals. Those efforts included meetings with other top American and British policy makers, several of whom offered support for Bohr's proposals.[8] The heart of Bohr's proposal was the control of atomic power through an organization similar to the League of Nations or the then-nascent United Nations. Bohr offered few details on his proposed international control system, beyond arguing that compliance should be based not on the threat of sanctions, but rather on giving international inspectors open access to nuclear facilities in all nations. Bohr advocated the complete exchange of all technical information on military and industrial atomic research, all of which should be subject to inspection. Such information would flow freely across national borders, preventing nations from developing atomic weapons in secret and protecting against the diversion of fissionable materials to military projects. Every site conducting research on any aspect of atomic energy anywhere in the world would be open to inspections by an international security agency. "In principle," J. Robert Oppenheimer, the wartime director of the Los Alamos laboratory, later wrote of Bohr's proposal, "everything that might be a threat to the security of the world would have to be open to the world."[9]

Bohr's renown as a physicist opened doors for him at the highest levels of power in London and Washington but failed to persuade Roosevelt and Churchill of his plan's merits. However, Bohr's ideas did resonate with such key figures in the Manhattan Project as Vannevar Bush, the director of the Office of Scientific Research and Development—which coordinated virtually all wartime R&D, including the Manhattan Project—and James Conant, a chemist and former president of Harvard University who worked closely with Bush. Other scientists at several Manhattan Project sites embraced Bohr's ideas or developed similar ideas independently. The scientists working at Los Alamos and most notably at Chicago's Metallurgical Laboratory, including the outspoken Hungarian-born physicist Leo Szilard, developed international control proposals during the war that were quite similar to Bohr's proposal.[10] Other wartime statements by Manhattan Project scientists, including the Franck Report, embraced the concept of international control.[11]

Civilian vs. Military Control

The debate about control of atomic weapons, like everything else, was transformed by Hiroshima and Nagasaki. In the weeks following Japan's subsequent surrender, the American people learned for the first time of the Manhattan Project and the scientists who developed the bomb. The end of the war and the eventual lifting of many wartime security restrictions allowed those scientists not only to openly discuss the bomb's implications with one another, but also for the first time to raise such issues with the public. They embraced the opportunity with relish. At the Met Lab in Chicago, scientists who had started to discuss the idea of a postwar organization even before Hiroshima quickly formed the Atomic Scientists of Chicago. Communication between the Manhattan Project sites was still restricted, but similar groups were established at Los Alamos and the Clinton Labs at Oak Ridge. While the Hanford Site saw less political activity in the immediate postwar period, groups of scientists were also soon established in New York, Philadelphia, and in such university communities as Cambridge, Berkeley, and Ann Arbor. In November 1945, these groups formed the Federation of Atomic (later American) Scientists (FAS). Though an all-volunteer group, the FAS charged dues, published newsletters, sponsored meetings, and maintained an active presence in Washington.[12]

Many scientists believed international control was the most important issue confronting the postwar world, but political realities forced them to focus first on the issue of domestic control of atomic energy. Soon after the FAS was created, Congress began consideration of the May-Johnson bill, which proposed a system for the domestic control of atomic energy that could have placed atomic weapons and research largely under control of the military. The scientists quickly organized to oppose May-Johnson, criticizing its secrecy provisions, which they believed would stifle scientific research, and the strong role it would give the military over atomic energy policy.[13] They offered testimony in Congress, wrote opinion pieces in newspapers, and hosted public meetings across the country. Their efforts were effective: an alternative to May-Johnson, the McMahon bill, was introduced in December, passed Congress in late July 1946, and signed into law as the Atomic Energy Act of 1946 on August 1.[14] The final bill created a civilian Atomic Energy Commission (AEC) without military members, placed some restrictions on the development of atomic weapons, and left open the door for negotiation of an international control accord.

The passage of the McMahon bill was a major political triumph for the scientists—though it may have been something of a Pyrrhic victory, since the final bill shared many similarities to the May-Johnson bill in terms of its security provisions. As historian Gregg Herken has noted, the political compromises needed to secure passage of the McMahon bill "meant that the principal aims of May-Johnson supporters were ultimately achieved." Alice Kimball Smith said the final law was a "qualified triumph" for the scientists and their political allies.[15] But the "scientist-activists" had proven to be exceedingly effective lobbyists with "considerable savvy in the techniques of public relations," as Paul Boyer has written.[16] They were able to speak with relative unity on the issue of domestic control, and their views carried weight because they alone understood the science behind the bomb's development in these early months of the atomic age. Their lobbying played a vitally important role in establishing the postwar structure for the management of U.S. atomic energy policy—a structure premised on civilian control that has evolved but still serves as the foundation for such policies today.[17]

STALEMATE AT THE UNITED NATIONS

The scientists hoped to build on their success in the domestic control debate when the focus shifted to international control during the development of the U.S. proposal for the first meeting of the United Nations Atomic Energy Commission (UNAEC) in June 1946. As part of that effort, Secretary of State James Byrnes formed an advisory committee to prepare a report for presentation to the UNAEC. The committee was led by Under Secretary of State Dean Acheson and Tennessee Valley Authority Chairman David Lilienthal, although much of the committee's report was written by Robert Oppenheimer, an advocate for Niels Bohr's ideas on international control.

The Acheson-Lilienthal report captured the spirit and incorporated the essential points of the approach to international control advocated by Bohr and others, most notably in its call for creation of an international body, the Atomic Development Authority (ADA), to promote the peaceful development of atomic energy. The ADA would have ensured compliance through inspections of atomic facilities and monitoring of raw materials rather than a system of clearly defined sanctions. The ADA would also have guaranteed the free exchange of scientific ideas, since the free movement of scientists and information would be essential to ensuring compliance while promoting international cooperation.[18] Acheson-Lilienthal did not become the U.S. proposal to the UNAEC, however, as Byrnes chose instead to advance a plan developed by Bernard Baruch, a politically influential financier whom Byrnes had selected to be the U.S. representative to the commission. The Baruch and Acheson-Lilienthal plans shared many key goals, but they differed sharply on how to reach those goals. In particular, the Baruch plan emphasized compliance through sanctions, not free scientific interchange, and rejected any early trust-building exchange of information, ensuring that the U.S would maintain its nuclear monopoly until the international control system was operating. As Paul Boyer has written, the Soviets viewed the Baruch plan as "a formula for perpetuating American nuclear superiority into an indefinite future."[19] The Soviets rejected Baruch's proposals, and hopes for international control quickly faded. The UNAEC continued to work on the issue for a few years, issuing reports that concluded the idea of international control was technically feasible and that an international agreement could help advance the development of atomic energy. But rising Cold War tensions ensured that the UNAEC would never reach agreement on the issues. The UN dissolved the commission in 1952, after the successful Soviet nuclear test in late 1949, but hopes for international control probably died with the Soviet rejection of the Baruch plan five years earlier.

THE SCIENTISTS AND INTERNATIONALISM

One cannot say with any certainty whether the international control of atomic weapons was viable in light of the international political tensions that transformed the world in the immediate postwar years. Historians have long debated the reasons for and implications of decisions by American and British political leaders on the development, use, and control of the atomic bomb, and whether the nuclear arms race between the U.S.

and the Soviet Union could have been averted if the scientists' proposals had been adopted. Many of the most politically active atomic scientists had even broader goals than international control, however, believing that the development of the atomic bomb both necessitated and made possible a fundamental change in the nature of international relations. They believed that the atomic bomb brought about such a quantitative increase in military power that it constituted a "qualitative transformation of the political environment," as international relations theorist Hans Morgenthau described the scientists' argument.[20] The common threat posed by the atomic bomb would underscore shared interests that transcended national boundaries, inevitably pushing nations toward world government.

These ideas recalled the hopes of Woodrow Wilson and the internationalist peace activists of the interwar era, who believed the world had been so repulsed by the horrors of World War I that statesmen would set aside traditional power politics and embrace international law and organization. Those hopes were embodied in the League of Nations and the Kellogg-Briand Pact, intellectual precursors to the Wilsonian international control regime envisioned by the Manhattan Project scientists. The Kellogg-Briand Pact clearly failed in its effort to outlaw war, though legal scholars Oona Hathaway and Scott Shapiro argue that it did begin the process of creating a world order in which aggressive conflicts between states have been far less common than prior to World War II.[21] The scientists' ideas also looked ahead to that new postwar world order. Ironically, the scientists' far-reaching hopes may have limited the appeal of their arguments. Political leaders did not believe that the development of the atomic bomb had transformed the nature of global politics, arguing instead that the pursuit of national interest would remain the driving force even in the nuclear age. As the influential theologian and political commentator Reinhold Niebuhr said at a conference in Chicago just a month after the war ended, a transformation of power politics was neither probable nor possible "even if you add the fear of almost complete mutual destruction to the other incentives which mankind has to overcome war."[22] This argument was embraced during and after World War II by the political leaders whose decisions sealed the fate of the scientists' proposals.

And yet, the scientists' international control proposals continued to resonate for years after Alamogordo. President Eisenhower's "Atoms for Peace" address in late 1953 owed a clear intellectual debt to them, combining warnings about the dangers of nuclear weapons with hopes for the peaceful use of nuclear power. That speech led within a few years to the creation of the International Atomic Energy Agency (IAEA). Decades later, the early years of the post-Cold War era were filled with talk of a "new Baruch plan" for controlling atomic energy, with hopes that the ideas first voiced by the scientists could have a better chance of being adopted after the collapse of the Soviet Union ended the rivalry that had doomed the original Baruch Plan. Arms control expert Randy Rydell notes that the Baruch plan was the precursor of the IAEA's nuclear inspection and verification schemes that remain central to the nuclear debate today.[23] While a new Baruch plan never emerged, the world community even today debates ideas that can be traced to the proposals of the atomic scientists. The United States and post-Cold War Russia have cooperated on fissile materials, and the United States has proposed a treaty to prohibit the production of highly enriched uranium and plutonium—concepts built on ideas first proposed by the atomic scientists. Echoes of those ideas can be heard in President Obama's 2009 Prague address outlining a vision for controlling nuclear weapons while supporting the peaceful uses of nuclear energy. The current debate about the international nuclear agreement with Iran raises arguments related to compliance (sanctions or inspections?) and the use of technical remedies to address political problems that were first debated by the atomic scientists 75 years ago. In a world of growing uncertainty, the ideas for an international control regime first voiced by the atomic scientists continue to resonate.

THE SCIENTISTS AND A POLITICALLY ACTIVE SCIENTIFIC COMMUNITY

The scientific community in the United States had little history of political engagement before World War II. The leading scientific professional societies, such as the American Association for the Advancement of Science founded in 1848, were created largely to promote science as a profession. Some more narrowly focused professional societies representing particular professional disciplines engaged in political debate, but only on issues that directly affected their members. During the Great Depression, groups like the American Association of Scientific Workers (AAScW) explored broader social and political issues, but they were relatively rare and did not have broad political reach.[24] The atomic scientists, in contrast, jumped whole-heartedly into the world of politics. They were labelled the "reluctant lobby," but many of them did not seem that reluctant. They lobbied members of Congress directly, testified at congressional hearings, wrote reports and opinion pieces, published newsletters, and organized conferences and rallies. While previous political efforts by scientific groups had tended to focus on influencing only elites, and only on issues affecting the scientific profession, the atomic scientists sought to influence the general public as well as elites in Congress, the Truman administration, academia, and the media. They showed a willingness to engage in grass-roots politicking in a manner far more reminiscent of a ward boss than a group of physicists and chemists. They engaged in coalition-building, reaching out to peace groups, religious organizations, labor unions, and other political organizations through the National Committee on Atomic Information (NCAI), which was formed to disseminate information in support of domestic control. This was a political movement unlike any before it, "a new current of thought and action among American scientists," as Edward Shils wrote a decade later in the *Bulletin of the Atomic Scientists*.[25] It helped shape the emergence of a more politically active scientific community in the postwar era.

The emergence of such a community was in many ways an inevitable consequence of the successful application of science to the development of new military technologies by the Manhattan Project and other American wartime endeavors. Those who worked on the Manhattan Project came face-to-face with the consequences of their research because, in the development of the atomic bomb, many of the same scientists carried the project from the blackboard to the test site at Alamogordo. As a result, they were not shielded by the time interval that had traditionally isolated scientists from the military consequences of their scientific labors.[26] The Second World War also fundamentally changed American attitudes about the relationship between scientific and technological leadership and national military and political power. Wartime scientific achievements helped open an era in which the United States, driven by the competitive pressure of the Cold War, engaged in a relentless mobilization by pouring massive resources into the *peacetime* development of the newest military technologies. American policy makers and the American public became obsessed after World War II with maintaining a technological edge over the Soviet Union, believing that advantage in the development of the newest military technology could, in the words of Don K. Price, provide "the crucial advantage in the issue of power" in the Cold War.[27] American scientists believed that their role in developing those technologies gave them both the technical expertise and the moral responsibility to undertake efforts to influence the policies that would govern their use. Yet such policy debates, while clearly affected by technical and scientific consideration, were fundamentally not about science but rather about such issues as the economy and national security. Beginning with the debate over controlling atomic weapons, the American scientific community became politically active on a wide variety of issues that had little to do with its traditional policy concerns, such as federal R&D spending.

The issue of controlling atomic weapons primarily involved broad questions of American foreign and defense policy—areas in which American scientists had never in the past claimed anything more than an amateur's interest. Yet the scientists who worked on the Manhattan Project claimed that their training and experience as scientists gave them a right to participate in debates about such questions. This constituted a fundamental change in the attitude of the American scientific community and provided the intellectual basis for subsequent political activism by American scientists on a wide variety of issues including the environment and, most notably, climate change.

The atomic scientists were, for a relatively brief time, at the center of American political debate. During the period immediately following Hiroshima and Nagasaki, the country was naturally gripped by the story of the bomb's development and its unprecedented destructive power. As a result, the atomic scientists were soon in the spotlight, as Americans tried to comprehend the bomb and what it meant for the postwar world. The scientists seized that moment, exerting real influence during congressional consideration of the May-Johnson and McMahon bills. Their expertise and opinions were sought by lawmakers and the public alike. "In the immediate wake of the war," historian Brian Balogh wrote, "the Manhattan Project scientists enjoyed extraordinary personal and collective prestige."[28] But their influence began to fade after Congress passed the McMahon bill and President Truman signed it into law. The scientists exerted less influence during the international control debate at the United Nations, where Bernard Baruch used his personal influence, connections, and political experience to shape the U.S. proposal to the UNAEC. Internal divisions emerged between those scientists who believed strongly that development of the bomb both necessitated and made possible the emergence of world government, and those who harbored less lofty goals.[29] The American public, once captivated by the scientists and their successful development of the atomic bomb, grew more wary of both science and the destructive powers that it had unleashed.

The increase in Cold War tensions soon dealt another blow, as questions were raised about the scientists' commitment to national security. The security hearings focused on Robert Oppenheimer are best known, but other scientists were subject to similar scrutiny as their arguments in support of the free flow of scientific information or the need for world government were increasingly called in question. Historian Paul Rubinson notes that the atomic scientists "fell out of favor as paranoia replaced postwar optimism" during the late 1940s and 1950s.[30] Some scientists also became increasingly intertwined with the national security state during the Cold War years, as the national laboratories, the Department of Defense and later the Department of Energy, and other federal agencies became important funders of scientific research—much of which came with strong security restrictions. Scientists became key players in what Brian Balogh calls the "prominstrative" state of pragmatic expert-driven policy making.[31] The atomic scientists' movement, which was briefly at the center of a national debate on issues of paramount importance, soon faded from the headlines.

And yet the FAS continued its work, and the *Bulletin of the Atomic Scientists* continued to publish and adjust its famous Doomsday Clock. The *Bulletin* added the phrase "a magazine of science and public affairs" to its masthead. "What was first conceived as an emergency, in which a few specific problems required resolution, established itself as a chronic condition," Edward Shils wrote in 1957, reflecting on a decade of political activism by the atomic scientists. Shils went on to note that such activism soon encompassed "areas of social life which might have appeared earlier to be unconnected with the interests of responsible scientists."[32] While the FAS continued to devote itself to issues related to nuclear weapons, such as efforts to ban atmospheric testing under the Partial Test Ban Treaty, the group started to expand its efforts during the early 1970s to broader energy, environmental, and social issues. The FAS opened a Washington office with a full-time staff and engaged actively in lobbying.[33] It was joined by other groups of politically active scientists, including those associated with Ralph Nader and the Center for Science and the Public Interest, and such groups as the Union of Concerned Scientists (UCS). The *Bulletin of Atomic Scientists* expanded its focus as well, devoting a growing amount of time to such issues as climate change. The *Bulletin* published a cover story on global warming in 1978 at a time when the issue received little attention from Congress or the American public, and climate change remains a major focus of the *Bulletin* today.[34] The FAS, UCS, and similar groups have remained active in lobbying Congress and the public while also pursuing legal strategies in the courts and regulatory agencies, their numbers expanding dramatically in recent decades as scientists became permanent fixtures on the Washington lobbying and think tank scene.[35] The origins of this politically engaged scientific community can clearly be found in the work of the Manhattan Project's "scientist-activists."

From the Manhattan Project to the March for Science

The election of Donald Trump has driven the scientific community to a new level of political activism, one perhaps not seen since the heyday of the atomic scientists' movement over 70 years ago.[36] Much of this activism is focused on traditional issues of science policy: fighting proposed cuts in federal R&D funding, supporting the use of evidence-based studies in shaping public policy, advocating for agencies and offices to bring scientific advice into the policy-making process. But much of it is also focused on broader issues of public policy, including the environment, climate change, privacy, and immigration. Scientists are forming political action committees to support their fellow scientists who want to run for office, writing letters to lawmakers, and hosting public meetings to ensure that the Trump administration funds scientific research and relies on scientifically sound studies to inform federal policies and regulations.[37] "An activist role is not an easy fit for many scientists," Amy Harmon and Henry Fountain wrote in the *New York Times*, noting that many scientists believe that they should merely present data without offering political interpretation. Ed Yong added in the *Atlantic* that these arguments are nothing new, since many in the scientific community have long believed that "scientists who engage in political advocacy are jeopardizing their credibility as objective, impartial, rational chroniclers of evidence."[38]

The Manhattan Project scientists who fought for control of atomic weapons rejected that argument, believing that they had an obligation as scientists to engage in political debate. They applied their expertise and experiences in the sciences to the debate of public policy issues that were not fundamentally about science alone, but rather about how science and technology affect national security, the economy, and, indeed, American society. The scientists who took to the streets during the March for Science were walking in the footsteps of those atomic scientists.

Notes

1. Nicholas St. Fleur, "Scientists, Feeling Under Siege, March Against Trump Policies," *New York Times*, April 22, 2017.

2. Some of the most notable works on the scientists' movement and the debate over control of atomic weapons include: Alice Kimball Smith, *A Peril and a Hope: The Scientists' Movement in America, 1945-47* (Chicago: University of Chicago Press, 1965); Paul Boyer, *By the Bomb's Early Light* (Chapel Hill, NC: University of North Carolina Press, 1994; originally published in 1985); Martin J. Sherwin, *A World Destroyed: The Atomic Bomb and the Grand Alliance* (New York: Vintage, 1977); and Richard Rhodes, *The Making of the Atomic Bomb* (New York: Simon and Schuster, 1986).

3. From a letter written in 1862; quoted in William S. McFeely, "The Letters of a World Watcher," *New York Times*, March 6, 1983.

4. William H. McNeil, *The Pursuit of Power* (Chicago: University of Chicago Press, 1982), 360

5. Niels Bohr, "Science and Civilization," *Times* (London), August 11, 1945.

6. Victor Weisskopf oral history interview, Niels Bohr Library, Center for the History of Physics, American Institute of Physics, New York, 29.

7. J. Rud Nielson, "Memories of Niels Bohr," *Physics Today* 16 (October 1963), 28–29.

8. The story of Bohr's efforts is well told in Ruth Moore, *Niels Bohr: The Man, His Science, and the World They Changed* (New York: Knopf, 1966); Sherwin, *A World Destroyed*; and S. Rozental, ed., *Niels Bohr: His Life and Work as Seen by His Friends and Colleagues*

(New York: John Wiley & Sons, 1967). Many of Bohr's letters, memorandums, and articles on international control can be found in the papers of J. Robert Oppenheimer, Manuscripts Division, Library of Congress, Washington. The author also wrote about Bohr's efforts in "The Physicist and the Politicians: Niels Bohr and the International Control of Atomic Weapons," in Reed M. Davis, ed., *Moral Reasoning and Statecraft* (Lanham, MD: University Press of America, 1988), 105–131.

9. J. Robert Oppenheimer, "Niels Bohr and Atomic Weapons," *The New York Review of Books* (December 14, 1964), 6.

10. Leo Szilard drafted a memorandum on international control for President Roosevelt that echoed many of the themes raised by Bohr, though Roosevelt did not see it before his death in April 1945. See Leo Szilard, "Atomic Bombs and the Postwar Position of the United States in the World," in Spencer Weart and Gertrud Weiss Szilard, eds., *Leo Szilard: His Version of the Facts* (Cambridge, MA: MIT Press, 1978), Document 101.

11. The Franck Report, written by a group of Met Lab scientists in June 1945, argued against the unannounced use of an atomic bomb against Japan, but that argument was premised in part on the assertion that "much more favorable conditions" for establishing an international control regime would be created by a demonstration of the bomb's power rather than its unannounced use against Japan. See "Political and Social Problems" (The Franck Report), June 11, 1945, File 76 (Interim Committee – Scientific Panel), Harrison-Bundy Files Related to the Development of the Atomic Bomb, 1942–1946, Manhattan Engineer District Files, Record Group 77, National Archives, Washington. The Franck Report was first published as Appendix B in Smith, *A Peril and a Hope*.

12. While the most prominent members of the atomic scientists' movement were physicists, it also included chemists and others who had worked on the bomb project. The story of the FAS and the groups at the Manhattan Project sites is well told in Smith, *A Peril and a Hope*. See also "The Atomic Scientists of Chicago," *Bulletin of the Atomic Scientists* (December 10, 1945); John A. Simpson, "They Did Something About the Bomb," *University of Chicago Magazine* (November 1946); an unsigned history of the Association of Los Alamos Scientists, October 24, 1945, Folder 5, Box II, Papers of the Association of Los Alamos Scientists, Department of Special Collections, University of Chicago Library, Chicago; and memorandums about "amalgamation" of the three scientists groups at Oak Ridge's Clinton Labs in Folder 3, Box I, Papers of the Association of Oak Ridge Engineers and Scientists and Related Groups (AORES papers), Department of Special Collections, University of Chicago Library, Chicago. The University of Chicago Library holds an extensive collection of documents about the scientists' movement.

13. Manhattan Project scientists pushed against stringent security measures even during the war, arguing that they were slowing progress in developing the bomb. Such concerns were particularly strong at the Met Lab in Chicago, where Leo Szilard and others were openly critical of the intense wartime security and "compartmentalization" of information. Such concerns clearly contributed to the emergence of a politically engaged scientists' movement in Chicago and were a driver in the writing of the Franck report. See, for example, Matt Price, "Roots of Dissent: The Chicago Met Lab and the Origins of the Franck Report," *Isis* 86 (June 1995), 222–244.

14. The domestic control debate is described in the official history of the U.S. Atomic Energy Commission, Richard Hewlett and Oscar Anderson, *The New World* (University Park, PA: Pennsylvania State University Press, 1962), 513–30.

15. Gregg Herken, *The Winning Weapon: The Atomic Bomb in the Cold War, 1945–1950* (New York: Vintage, 1982), 148; Smith, *A Peril and a Hope*, 325.

16. Boyer, *By the Bomb's Early Light*, 62.

17. The AEC existed until 1974, when the Energy Reorganization Act shifted its regulatory powers to a new Nuclear Regulatory Commission. Programs promoting the development of nuclear energy were shifted to the Department of Energy, which was created in 1977. This basic structure remains in place today.

18. U.S. Department of State, Committee on Atomic Energy, *A Report on the International Control of Atomic Energy* (the Acheson-Lilienthal report), March 16, 1946, ix. Detailed accounts of the drafting of the Acheson-Lilienthal report and the UNAEC debates can be found in Hewlett and Anderson, *The New World*, 531–54 and 582–619. See also Herken, *The Winning Weapon*, 151–91.

19. Boyer, *By the Bomb's Early Light*, 55. As Boyer notes, the differences between the Baruch and Acheson-Lillienthal plans are not surprising, since the Baruch plan was written by a group drawn far more heavily from Wall Street than the scientific community.

20. Hans J. Morgenthau, "Decision-making in the Nuclear Age," *Bulletin of the Atomic Scientists* (December 1962), 7.

21. Oona A. Hathaway and Scott J. Shapiro, *The Internationalists: How a Radical Plan to Outlaw War Remade the World* (New York: Simon & Schuster, 2017).

22. "Proceedings of an Atomic Energy Control Conference at the University of Chicago," September 19–22, 1945, Addenda, Folders 6 and 7, Box IV, Papers of the Atomic Scientists of Chicago, Department of Special Collections, University of Chicago Library, Chicago.

23. Randy Rydell, "Going for Baruch: The Nuclear Plan That Refused to Go Away," *Arms Control Today*, 36: June 2006, 45–48. Thanks to Zachary Davis of the Center for Global Security Research at Lawrence Livermore National Laboratory for his thoughts on the relevance of the atomic scientists' ideas from the Atoms for Peace proposal to today's debates about Iran and efforts to control nuclear materials.

24. As Edward Shils notes, the AAScW also offered a "radical, more or less Stalinist, criticism of American society" (Edward Shils, "Freedom and Influence: Observations on the Scientists' Movement in the United States," *Bulletin of the Atomic Scientists* (January 1957, 13). See also Jessica Wang, "Scientists and the Problem of the Public in Cold War America, 1945–1960," *Osiris* (2002), 323–347.

25. "The Reluctant Lobby," *Newsweek* XXVI (December 3, 1945), 42; Shils, "Freedom and Influence," 13; "National Committee on Atomic Information," *Bulletin of the Atomic Scientists* 1 (December 24, 1945); Smith, *A Peril and a Hope*, 227–28. The work of the NCAI is detailed in letters written by Daniel Melcher, the NCAI's first director, which

can be found in Box 18, records of the National Committee on Atomic Information, Manuscripts Division, Library of Congress, Washington.

26. These ideas are emphasized in Simpson, "They Did Something About the Bomb," 3; Henshaw, "Science and Public Policy as Viewed by the Oak Ridge Scientists and Clinton Laboratories," November 7, 1945, Folder 3, Box II, AORES papers.

27. Don K. Price, "The Republican Revolution," in William R. Nelson, ed., *The Politics of Science* (New York: Oxford University, 1968), 5.

28. Brian Balogh, *Chain Reaction: Expert Debate and Public Participation in American Commercial Nuclear Power, 1945–1975* (New York: Cambridge University Press, 1991), 34.

29. This became clear in a rift between the FAS and the Emergency Committee of Atomic Scientists (ECAS). Albert Einstein, Leo Szilard, and other members of ECAS remained ardent advocates of world government even as Cold War tensions rose in late 1946 and 1947, but leading members of the FAS argued the scientists needed to recognize those changes in global politics. See, for example, a letter from Hans Bethe to Joseph Halle Schaffner, April 29, 1947, Folder 15, Box 1, Papers of the Emergency Committee of Atomic Scientists, Department of Special Collections, University of Chicago Library, Chicago.

30. Paul Rubinson, *Redefining Science: Scientists, the National Security State, and Nuclear Weapons in Cold War America* (Amherst, MA: University of Massachusetts Press, 2016), 19.

31. Balogh develops this concept in *Chain Reaction*.

32. Shils, "Freedom and Influence," 13.

33. Robert Smith, "The New Scientist-Advocates," *Bulletin of the Atomic Scientists*, February 1975, 16–18.

34. David Kaiser and Benjamin Wilson, "American scientists as public citizens: 70 years of the *Bulletin of the Atomic Scientists*," *Bulletin of the Atomic Scientists*, January 2015, 20–21.

35. Scientists are also active in lobbying and interest groups founded and led by nonscientists, such as the Natural Resources Defense Council.

36. The scientific community did briefly capture the nation's attention during the 1980s and the debate over "nuclear winter," when Carl Sagan in particular drove a national debate over the risks of nuclear winter following a large nuclear exchange. See Lawrence Badash, *A Nuclear Winter's Tale: Science and Politics in the 1980s* (Cambridge, MA: The MIT Press, 2009). Jill Lepore notes the links between the current debate about climate change and the debate about nuclear winter, during which some of the institutions that question climate science were established. See Jill Lapore, "The Atomic Origins of Climate Science," *New Yorker*, January 30, 2017.

37. Ed Yong, "Professor Smith Goes to Washington," *Atlantic*, January 25, 2017.

38. Amy Harmon and Henry Fountain, "In Age of Trump, Scientists Show Signs of a Political Pulse," *New York Times*, February 7, 2017, D1; Ed Yong, "Do Scientists Lose Credibility When They Become Political," *Atlantic*, February 28, 2017.

SECTION III

FACTS AND FICTIONS

CHAPTER SEVEN

Pursuing the Cancellation of the Apocalypse

Terry Tempest Williams' *Refuge* and Rebecca Solnit's *Savage Dreams*

Daisy Henwood

Writing about the aftermath of the Manhattan Project, Bruce Cameron Reed states, "these legacies will be with us for decades to come. They include America's post-war military power and political influence, the enormously expensive Cold War, the thousands of nuclear weapons still extant today, ongoing weapons development programs in potentially unstable countries, the threat of nuclear terrorism, costs of remediating environmental damage at weapons-production and testing sites, and public apprehension with nuclear energy."[1] Reed rightfully observes that 'the Bomb' altered the global political arena, yet he does not contend with the social impacts of the development, testing, and use of nuclear weapons. Focusing on the "local" impact of weapons development and considering nuclear testing to be a legacy of the Manhattan Project in both a physical and ideological sense, this essay discusses the impact of bomb testing on life in the American desert. Discussing Terry Tempest Williams' *Refuge: An Unnatural History of Family and Place* (1991) and Rebecca Solnit's *Savage Dreams: A Journey into the Landscape Wars of the American West* (1994), I consider the ways these writers each depict nuclear testing, fallout, and waste, and envision a future shaped not by nuclear destruction, but by activism.

Savage Dreams offers a comprehensive insight into antinuclear writing, as well as a nuanced portrayal of the effects of nuclear testing and

133

its roots in the Manhattan Project itself. Solnit's text is rarely discussed, and thus provides a unique opportunity not only to trace the effects of nuclear testing in the desert west, but also to unpack the very representation of those effects. Similarly, *Refuge* is rarely discussed within a nuclear context—indeed, most of the text itself avoids mentioning nuclear testing altogether. I use it here, then, as a comparative text in order to illustrate not only the widespread effects of nuclear testing in the Nevada desert, but also to illustrate the very evasiveness by which nuclear effects have been and continue to be characterized. Both texts display radical approaches to the desert west that take into account its beauty and its danger, its status as home-place and toxic dump. And, although published in the 1990s, both texts contend with nuclear futures of which we are still unsure.

Daniel Cordle observes that, "after 1945, nuclear technology became the fabric out of which everyday experience was woven. Its possible effects had to be assimilated, even if for most people, most of the time, this involved acceptance or denial of, not active resistance to, this new condition of reality."[2] This new condition of reality is the (mostly) invisible legacy of "the Bomb." Williams and Solnit write about this new reality in terms of physical fallout—in the desert's dust and its effect on the body—and as an emotional or affective fallout that leads to their involvement in antinuclear protest. In *Savage Dreams*, Solnit describes protests at the Nevada Test Site as well as her interactions with indigenous groups and local residents. For Solnit, either accepting or denying the legacies of nuclear weapons becomes impossible as she confronts the devastating physical and emotional effects of testing. Nuclear technology also underlies *Refuge*, as throughout the text Williams connects her family's proximity to the test site to their high cancer rates. For Williams, too, acceptance and denial are not options. Both women choose a third path: their texts culminate in acts of resistance.

This resistance comes from the desire of both women to confront unstable, uncertain nuclear futures. Using Rob Nixon's concept of "slow violence"—"violence that occurs gradually and out of sight, a violence of delayed destruction that is dispersed across time and space, an attritional violence that is typically not viewed as violence at all"—to describe the way various nuclear fallouts act in both texts, I trace an apocalyptic narrative in *Refuge* and *Savage Dreams* that starts with above-ground nuclear testing in the 1950s, moves through the "present" of each text in the late eighties, and ends, if it ends at all, in the very distant future.[3] Through the slow understanding of the long-term effects of nuclear technology, both women write their way into a nuclear future that avoids or extends beyond the immediate apocalypse symbolized by above-ground testing, and the slow apocalypse engendered by an increasingly toxic and toxified land and population. Apocalypse, for James Berger, works in three ways: "First, it is the eschaton, the actual imagined end of the world…Second, apocalypse refers to catastrophes that resemble the imagined final ending …an end of something, a way of life or thinking…[Apocalypse], finally, has an interpretive, explanatory function."[4] Reading Berger's three definitions of apocalypse in terms of the past, present, and future of each text, I split my analysis into three sections. The first, "1957," centers on the immediate impacts of nuclear testing in each text, discussing Williams' memory of a nuclear detonation and Solnit's conversation with a woman whose brother died of radiation exposure. The second, "1987," reframes these narratives in terms of what I call the "slow apocalypse," discussing the bodily effects of exposure in both texts as a form of Berger's "end of something, a way of life or thinking." The final section, "10,087," plays on the projected date by which nuclear waste storage is supposedly no longer a problem. Focusing on how each writer imagines a future in the nuclear world, this section uses Berger's notion of apocalypse as an interpretive tool to gesture to the ways each woman envisions a future beyond, or instead of, apocalyptic endings. I close by discussing how, for both women, this future is marked—even created—not by death and destruction, but by activism and its attendant hope for the future.

Charles E. Gannon gestures toward a legacy that extends beyond the physical effects of nuclear technology when he asserts, "if detonated, a nuclear bomb may vaporize millions, but even by merely existing, it shatters our ability to affix limits, to grasp the world in our accustomed framework of the finite."[5] He points to an epistemological crisis, in which our ability to "know" is compromised by nuclear weapons' literal and figurative engagement with ruptures, with endings. Jacques Derrida writes, "what allows us perhaps to think the uniqueness of nuclear war, its being-for-the-first-time-and-perhaps-for-the-last-time, its absolute inventiveness, what prompts us to think even if it remains a decoy, a belief, a phantasmatic projection, is obviously the possibility of an irreversible destruction."[6] Derrida refers here to the representational crisis of the nuclear world; being able to think total nuclear destruction relies on us understanding that it means absolute, irreversible endings. Yet, even thinking in this way marks a similar kind of "decoy" to not thinking it at all, to turning a blind eye, and our understanding of the nuclear future is, as Derrida asserts, "phantasmatic projection"; we simply do not know what will happen. If the epistemological crisis of nuclear technology ruptures our ability to think or imagine a future, its legacy surpasses physiological effects. Thus, while Solnit and Williams document the effects of nuclear testing in the desert and on the body, equally important to both is a reclamation of representational power. For Nixon, the crisis of slow violence is often "representational"; he wonders "how to devise arresting stories, images, and symbols adequate to the pervasive but elusive violence of delayed effects."[7] Contending with nuclear fallout(s), Solnit and Williams work through the representational issues facing the nuclear world by refusing to ignore its various and continued effects on those living alongside the Nevada Test Site. And by revealing narratives previously ignored, or silenced, *Savage Dreams* and *Refuge* become sources as well as records of activism, seeking to (re)present a future for the nuclear world.

1957

The "eschaton" of which Berger writes is most evident in *Savage Dreams*. The text opens with Solnit's arrival at the Peace Camp outside the Nevada Test Site in the 1980s, where she hears stories of above-ground tests and learns about Los Alamos. These stories evoke the sense of doom Berger associates with his first definition, the "actual imagined end of the world," and Solnit begins to consider the apocalyptic nature of the detonations that took place at the test site between 1951 and 1963:

> Test is something of a misnomer when it comes to nuclear bombs. A test is controlled and contained, a preliminary to the thing itself, and though these nuclear bombs weren't being dropped on cities or strategic centers, they were full-scale explosions in the real world, with all the attendant effects. I think that rather than tests, the explosions at the Nevada Test Site were rehearsals, for a rehearsal may lack an audience but contains all the actions and actors. The physicists and bureaucrats

managing the U.S. side of the Arms Race had been rehearsing the end of the world out here, over and over again.[8]

"Test," for Solnit, is an inadequate characterization of the destructive power of the bombs detonated in Nevada, and in the very evasiveness of this term there lies the slow violence of the test site. These "tests" affect real people, in real time; they contain "all the actions and actors" of the show. Thus, while downwinders (communities downwind of the test site, most exposed to the dangers of nuclear fallout) are not "strategic centers" in terms of global politics, they represent a political decision in which certain groups—the Western Shoshone, Mormon communities, and rural ranchers—have been deemed worthy, or, rather, worthless, enough to be unwittingly subjected to radioactive fallout.[9] Framing this revelation in terms of the apocalypse, Solnit gestures toward the way "the end of the world" means, at once, global nuclear catastrophe, and the end of the individual worlds—families, homes, lives—of the downwinders. Solnit writes, "the Department of Energy's own handbook of nuclear tests lists [Hiroshima and Nagasaki] as tests. To call such an act a test clarifies how far the mindset of scientific control had warped the vision of those who would call all future bomb explosions tests, no matter what their effect on the world around them."[10] Even when the effects of the bomb are explicitly meant to destroy human life, "test" is the preferred, sanitized term. Joseph Masco argues, "nuclear explosions at the Nevada Test Site were not merely tests; they were the entire performance, communicating to the world the U.S. possession of, and commitment to, weapons of mass destruction."[11] Unpacking the carefully cultivated innocence of the term "testing," Masco, too, reframes these "tests" in terms of political performances played out on a global, and at the same time local, scale; both the "performances" Masco sees and the "rehearsals" Solnit describes signal a smoke-and-mirrors theatricality played out in the nuclear desert. The destructive force of the test site thus hides behind the very language of nuclear "testing" as a "safe" endeavor; its local toxicity is obscured by the showmanship of its global importance.

Refuge only directly contends with narratives of nuclear testing in the epilogue, in which Williams confirms that she witnessed a nuclear test as a child in 1957, and begins to connect the "flash of light in the night in the desert" with her mother's cancer.[12] The text displays the slow violence of nuclear testing by not explicitly detailing that violence until the very end;

it is gradual and out of sight. This end, like the apocalypse hidden under the "tests" in *Savage Dreams*, is obscured by the government's version of events. Williams explains, "most statistics tell us breast cancer is genetic, hereditary… What they don't say is living in Utah may be the greatest hazard of all."[13] Here, Williams draws attention to the "official" denial of the dangers of nuclear testing. She later writes, "again and again, the American public was told by its government, in spite of burns, blisters, and nausea, '…the tests may be conducted with adequate assurance of safety…' Assuaging public fears was simply a matter of public relations."[14] Again, governmental discourse obscures the effects of nuclear technology. Engaging with the representational legacies of nuclear testing—the vagueness of the "they," the derisive nod to strategies of "public relations"—Williams understands that the widely espoused narrative of nuclear safety is responsible for allowing the continuation of the test site's slow violence. Alan Nadel connects this strategy with a larger cultural narrative of acceptance and denial associated with nuclear technology, arguing, "the American cold war is a particularly useful example of the power of large cultural narratives to unify, codify, and contain—perhaps intimidate is the best word—the personal narratives of its population."[15] Intimidation tactics at the level of narrative, in the very information given to downwinders, are the determining factor in the slowness of the violence of the test site. That is, the denial of illness, of ill-effects, determines that the damaging by-products of nuclear testing are rendered unbelievable by the very people responsible for them, and thus delay the realization of the apocalyptic, the ending, effects of nuclear testing.[16] While the immediacy of the nuclear flash is undeniable, it is only belatedly, partially known—Williams herself is uncertain even decades later, having recalled the flash first in a dream. Writing about the patterns of slow violence she has encountered since she was a child, representing the effects that have been deliberately obscured, Williams counters the denial and invisibility proffered by the Atomic Energy Commission, by the naming of detonations "tests," and even by her own mind's repression of the nuclear experience.

In *Savage Dreams*, Solnit includes the narrative of a woman who, like Williams, lives in Utah. Janet Gordon tells Solnit that she grew up "150 miles east of the Nevada Test Site," and was "twelve years old when the testing started, in 1951."[17] Gordon, a Mormon farmer, speaks of her brother's exposure to nuclear fallout: "Kent came back into camp one evening and he was very sick, he had burns on his skin like a really severe sunburn, he was throwing up, he had a bad headache, and he wondered if it had anything to do with the test they'd set off over in Nevada."[18] Gordon's narrative, and her brother's symptoms, pertain to the simultaneous knowing and not-knowing involved in living downwind of the test site; they "wonder" whether these symptoms have anything to do with nuclear fallout, but make no explicit connection outside of the private, familial narrative they share among themselves, despite the visibility of Kent's illness and the absence of any other cause. Gordon recalls, "they knew there was a test, it had been on the radio."[19] In keeping with the test site's secrecy, normalizing testing by announcing it on the radio acts as an obscuring agent, a tool used by the government to disrupt connections tentatively drawn between illness and nuclear tests; if it's announced on the radio, it must be safe. Their narrative avoids the terms of the apocalypse by situating nuclear testing within the perpetual routine of the everyday.

1987

Gordon contends that, while no one was told what the effects of radiation were at the time, Kent's symptoms were "classic," and he "died by inches" of a fast-moving pancreatic cancer.[20] Robert J. Lifton describes the "general sense of A-bomb disease as a thing apart from ordinary medical problems—more obscure, devious, ubiquitous, in every way deadly."[21] Though Lifton is writing about Hiroshima, the insidiousness associated with radiation exposure is described in such terms on the test site, too, drawing a connection between the Manhattan Project's direct effects, and its ongoing local impacts. Illness slowly arising from the tests in the fifties thus shapes the depiction of nuclear legacies in both texts, for it is latently—in the appearance of Kent Gordon's and Diane Dixon Tempest's cancers—that the true effects of radiation exposure are revealed. Berger considers apocalypse in terms of "revelation, unveiling, uncovering."[22] If apocalypse refers to this revelation, the terms of its unveiling, as well as the effects it unveils, are indicative of slow violence, and contribute incrementally to Berger's second definition of apocalypse, an end to a way of thinking and knowing the world.

Thus, gradual realization in both texts takes on the character of a slow apocalypse; both take time to come to terms with the realities of nuclear testing in their present. Williams realizes the cause of her mother's (and grandmother's and aunt's) cancer at the end of *Refuge*, and Solnit witnesses the illness of others and wonders how her own presence at the test site might translate, belatedly, into a nuclear sickness. Berger argues the apocalypse is "semantically unsayable…translated into politically somatic forms, into symptoms on the body politic."[23] In *Refuge* and *Savage Dreams*, the apocalypse is literally somatic, taking place on the site of the body even decades after above-ground testing has stopped. Bypassing linguistic representation altogether, nuclear destruction, so hidden by the discourses used by the government, appears writ on the body in the form of cancers and symptoms of radiation sickness.

Cassandra Kircher notes *Refuge*'s "deliberate delay of the cause-and-effect relationship between nuclear contamination and cancer [that] mirrors the naiveté of a society that understood the immediate horror of atomic bombs but had yet to learn about their long-term effects."[24] Connecting her mother's illness to nuclear testing only at the end of the text suggests an ambivalence, on Williams' part, concerning the connection itself. As with Kent Gordon's sickness, the ambivalence here marks an epistemological paralysis in which Williams both knows and does not know, is both sure of and yet persuaded against the connection between her mother's illness and nuclear testing. Yet, as Kircher points out, the delay mimics the slow revelation of radiation's effects. In *Refuge*, the narrative of nuclear exposure is refracted through the narrative of illness; while nuclear testing closes the narrative, it is illness that opens it. Williams writes, "most of the women in my family are dead. Cancer. At thirty-four I became the matriarch of my family."[25] Cancer seems to be the dominant destructive force of this text, and the focus of Williams' preoccupation with the future. Here, after all, she details the destruction of a familial legacy—her inheritance, the position of matriarch, comes much earlier than she'd ever imagined due to a sickness characterized by its very unpredictability; the epistemological crisis engendered by nuclear technology is, for Williams, a characteristic of cancer, too.

Recognizing the connection between environmental and bodily damage, Heather Houser writes, "ecosickness narratives involve readers ethically in our collective bodily and environmental futures."[26] The government's narrative works to disrupt this connection; the affective charge of nuclear destruction has been displaced onto the body, rather than connected with it, as a direct result of the government's tactic of denying any absolute connection between the two—they do not admit that the Utah landscape is toxic. Invoking these collective bodily and environmental futures, Williams comprehends simultaneously the unintelligible presence of nuclear testing in everyday life, and the slow violence of cancer on the body. Williams describes her mother's cancer as something that "begins slowly and is largely hidden," echoing the ways the toxicity of the nuclear landscape is largely invisible.[27] The slowness also pertains to incremental toxicity; the cancer, slow itself, is born of the slow violence of prolonged exposure to fallout. Susan Sontag writes of cancer as "a disease that doesn't knock before it enters, …a ruthless, secret invasion," an analysis that speaks directly to the language of nuclear technology.[28] Kept secret, imagined as useful for a kind of invasion-at-a-distance, a nuclear bomb is the very definition of destruction that does not knock; it is, in the end, a sudden, deadly power.

Masco points out the "strange reliance Americans now have on nuclear threat to organize politics and experience," explaining, "so many Americans, from so many different social positions…understand…non-nuclear, non-military event[s], in decidedly nuclear terms."[29] The discourse of nuclear war is so embedded in the language of the American people that it appears universally applicable to moments of crisis. Throughout *Refuge*, the legacy of nuclear testing resides not only in its literal connection to Williams' mother's cancer, but also in the very language used to describe it. Williams writes, "we can surgically remove it. We can shrink it with radiation. We can poison it with drugs. Whatever we choose, though, we view the tumor as foreign, something outside ourselves. It is however, our own creation. The creation we fear."[30] Aside from invoking the use of radiation to shrink a disease caused by radiation, Williams' Frankensteinian characterization of cancer as something we, in our bodies, create and fear is akin to the fear and creation associated with nuclear weapons. Nuclear technology has thus far eschewed all responsibility for its ill effects, yet Williams' description of cancer, reframed in terms of an analysis of nuclear technology, forces the onus back onto nuclear weapons and the "physicists and bureaucrats" responsible for them.

Solnit also locates an epistemological crisis in the body. She writes, "I remembered to be afraid of the dust [at the Peace Camp], the dust that might be radioactive…the dust that might be the dust of hundreds of nuclear tests."[31] The dust seems innocuous. It only "might" be radioactive. As such, the dust becomes something Solnit must remember to fear. Describing the desert dust, Solnit engages with the same ambivalence as Williams, who concedes "[she] cannot prove that [her] mother, Diane Dixon Tempest, or [her] grandmothers, Lettie Romney Dixon and Kathryn Blackett Tempest, along with [her] aunts developed cancer from nuclear fallout in Utah," but also, pointedly, notes "[she] can't prove they didn't."[32] There is no way of knowing how or even whether the radioactive dust will affect you, us, them in the future, yet Williams uses the same evasiveness, the same hedging uncertainty as the government's discourse itself; ironizing their refusal to neither confirm, nor deny, she leaves us in no doubt where she stands. *Refuge* and *Savage Dreams* therefore both reflect Berger's assertion that "post-apocalyptic discourses try to say what cannot be said (in a strict epistemological sense)."[33] Solnit cannot say whether the dust will affect her, she does not know, but by voicing the fear, Solnit voices the silenced reality of the test site's danger.

Solnit repeatedly voices this fear, writing, "I don't know now whether coming to the Test Site will kill me, whether some small particle of strontium or cesium in the dust will inaugurate a course of growth that will prove fatal."[34] Cordle describes the Cold War as a period of "fraught stasis," and Solnit points here to the way not only global politics entered into this stasis, but that it also played out on the site of the individual body.[35] Not knowing whether you have been exposed to radioactive fallout, the body enters a static, suspended mode in which it is both sick and well. Moreover, the precautions protesters and workers can take against this are negligible at best. Solnit records the guidelines given to activists by American Peace Test: "There is little that can be done to protect your body from beta and gamma rays which are unseen and penetrate your body. Alpha particles, however, may have longer term effects. They are found on dust particles that can be breathed in or ingested. Cover your face when walking in the wind. Do not eat food dropped on the ground. Don't use bare, dirty hands for eating."[36] The dust is threatening, insidious. This advice is driven more by anxiety than by protective action; it exacerbates both stasis and fear. Yi-Fu Tuan defines anxiety as "a presentiment of danger when nothing in the immediate surroundings can be pinpointed as dangerous. The need for decisive action is checked by the lack of any specific, circumventable threat."[37] The dust in the Nevada desert morphs the landscape into the body, making the internal self an equally (possibly) threatening and inescapable place. Like the land, it is both toxic and innocuous. Solnit's anxiety, like the cancer in the bodies of *Refuge*, acts as another form of slow apocalypse, leaving the traces of nuclear technology not only on the land, or in the physical degradation of the body, but etched onto the psyche, into the knowledge in/of this place, too.

10,087

In line with Berger's third definition, the apocalypse ultimately becomes an interpretive tool with which Solnit and Williams work to imagine a future, to think their way beyond the present moment of fear. Lawrence Buell considers *Refuge* in terms of the unknown. Discussing the epistemological uncertainty Williams associates with illness, with the landscape, and with nuclear technology, he observes, "indeterminacy at the level of knowledge itself exercises a kind of determination as an act of imagination: ensconcing toxic anxiety as a psychological reality and as a cause of immiseration in a good part because of the inability to know."[38] This inability plays into worries about Williams' own place in the toxic future. Documenting her mother's illness, Williams wonders whether "cancer [is her] path, too."[39] In the epilogue, Williams writes, "I belong to the Clan of One-Breasted Women. My Mother, my grandmothers, and six aunts have all had mastectomies. Seven are dead...I've had my own problems: two biopsies for breast cancer and a small tumor between my ribs diagnosed as a 'borderline malignancy.'"[40] "This is my family history," Williams tells us.[41] Implicitly, it is also her family's future. Cancer, as a result of nuclear fallout, becomes the apocalyptic metaphor through which Williams comes to terms with her uncertain, unstable future. Like the language through which nuclear war is implied, the notion of cancer as future—uncertain or not—represents a kind of fatalism at the end of *Refuge*. As John Beck points out, "once unleashed, the danger of nuclear destruction can never be removed."[42] If the damage can never be undone—if, as Reed asserts, the nuclear bomb can never be "un-invented"—what kind of future can Williams and her family expect?[43] For Williams, as for so many people caring for sick loved ones, or who are sick themselves, the future is immediate. It is closer—frightening, but in an imaginable if not understandable way.

In this sense, interpreting the apocalypse from the inside out, from her body out to the world, Williams sees a difficult path ahead, but it is a path that pushes her to think, to question, and to imagine her own place within the nuclear future, within the Clan of One-Breasted Women.

For Solnit, the nuclear future, in the form of nuclear waste, stretches for millennia. Describing the government employees at Livermore Lab, she reveals, "the physicists said that nuclear waste was not their department, but that of geologists, and so they didn't think about it."[44] Refusing the legacy of denial attached to this sentiment, refusing once again the bureaucratic discourse of evasion and avoidance, Solnit uses *Savage Dreams* to contemplate the effects of nuclear waste on the distant future. She traces a narrative legacy of the Manhattan Project, calling into question the language of evasion, of denial, continually employed in discourses surrounding nuclear technology. Centering around the unfathomably long half-life of plutonium, Solnit's discussion of the proposed nuclear waste storage strategies in the Nevada desert emphasizes the futility of attempting to prepare for 24,000 years of storing radioactive waste. Nuclear waste, in its ability to harm all life in both the present and the future, thus becomes indicative of a technoscientific disregard for both the present state of the earth, and its ability to sustain life into the future. There is no limit to this destruction; it takes millennia for nuclear waste to decay, and more waste is produced, and improperly stored, every day. The stuff of nightmares, nuclear waste comes to act in *Savage Dreams* as yet another indicator of the slow, apocalyptic, violence wrought by nuclear technology.

Highlighting once again the naivety and the willful ignorance of governmental claims, Solnit writes, "plutonium and other wastes weren't considered pollutants, since they didn't go into the environment—but where would they go?"[45] Here, Solnit locates within the narrative of nuclear waste the same disregard for consequences by which nuclear testing itself is so marked. With a lack of awareness of the ongoing dangers of nuclear waste—or perhaps a deliberate shying away from the facts of half-life and ineffective waste storage—the U.S. government's storage program has been consistently lacking. Solnit describes "rusting barrels in the ocean off San Francisco," and "leaking storage tanks in Hanford, Washington," all of which are "time-bomb monuments to the underestimates of the past."[46] For Solnit, the disregard for nuclear safety into the future is inextricable from the disregard for environmental and health implications in the present. Moreover, describing the future of this waste in the language of nuclear war—the current storage facilities are "time-bombs"—Solnit, like Williams, mirrors the inescapability of nuclear technology in the very language with which she describes it.

Detailing the planned Yucca Mountain nuclear waste storage facility, Solnit further emphasizes the incapability with which we are faced when dealing with nuclear materials, revealing, "as far as many scientists and activists are concerned, adequate storage is an idea that has not been realized yet, and may be unrealizable."[47] The inescapable fact is that the future of nuclear waste is unknown, unpredictable, and no one is equipped to deal with it effectively. For Masco, Yucca Mountain is where the Nevada Test Site "confronts its own apocalyptic excess and, in an effort to control that excess, is expanded—exponentially—to the point of self-contradiction and failure."[48] This failure, the futility of scientific prediction when it comes to nuclear waste, echoes more broadly Elizabeth Ammons' claim that "Western science practices the very alienated reasoning that has led us to believe that human beings can and should conquer and control nature."[49] Nuclear testing is thus embroiled in a paradox of its own creation, as its scientific aptitude, in terms of arms development, is its very ineptitude when it comes to waste storage. It is both progressing and prohibiting, and its faith in the control of nature—encapsulated by the faith that Yucca Mountain may be modified to contain millennia of radioactive waste—is an unequivocal failure, even an apocalyptic death wish.

Solnit explains that, due to the unpredictability of the nuclear future, the government agencies responsible for storing waste into the future have landed upon an arbitrary length of time for which they must be able to store the waste. The allocated 10,000 years covers less than half of the dangerous half-life of plutonium, and, even aside from coming up with an effective waste storage technology, the problem of communicating the toxicity of nuclear waste that far into the future is pressing. Solnit writes:

> The [Department of Energy] expects that in 10,000 years our language and culture will be extinct, since none has ever lasted a fraction of that time. Marking the waste-deposit sites in such a way that the warnings will last ten millennia and be meaningful to whomever may come along

then has been something of a challenge to the DOE's futurists. There were proposals … to establish a nuclear priesthood, which would hand down the sacred knowledge from generation to generation. Others proposed forbidding monuments of a vastness that would survive the erosion of all those years, though any monument could attract curiosity and no inscription was guaranteed to make sense.[50]

The ridiculousness of these suggestions correlates directly to the inability to know, think, or imagine 10,000 years into the future, and the suggestions made here are more reminiscent of a bad science fiction plot than a serious attempt to deal with the uncertainty of the nuclear future. Science has reached its epistemological limits; at this moment, it relies on religious narratives, a "priesthood," and in its turn away from rationality, towards faith, it ultimately undercuts its own "scientific" authority. For Nadel, nuclear technology "empowers an absurd discourse, one that necessarily elides the distinction between history and science fiction, for it necessitates the understanding of an event that cannot exist retrospectively."[51] It also, apparently, cannot exist preemptively. Solnit's slow apocalyptic future engenders not a grand renewal, but an acknowledgement of the decay and destruction connected to improper waste storage. Yet there is also something heartening in Solnit's recognition of the absurd. Understanding the difficulties of storing vast amounts of nuclear waste into the future, Solnit attempts to consider the unimaginable. Using the slow apocalypse of nuclear waste storage as an interpretive tool, Solnit sees, in the ridiculousness of prediction, the ridiculousness of the present state of nuclear technology. Absurdity here calls to attention the dangers of nuclear technology's flippancy, its casual, evasive, and covert destruction. Nixon writes, "the narrative imaginings of writer-activists may…offer us a different kind of witnessing: of sights unseen," to which, I would add, futures unknown.[52] These possible, imagined futures, then, are what drive the texts all along, and it is to these alternatives I shall turn finally.

Cancelling the Apocalypse

For Masco, "Nuclear ruins are never the end of the story in the United States but, rather, always offer a new beginning."[53] Solnit imagines a world of nuclear ruins when she describes Yucca Mountain. But rather than situating the new beginning this might engender at the end of the projected 10,000 years, she proposes a new beginning now. Solnit and Williams both contend with the difficulties of not-knowing, of trying to plan for a future that is unimaginable. Yet, in the end, their way out of it is to act. By even engaging with thinking about this unknowable future, both women act defiantly in the face of a technoscientific narrative that wants them to accept, comply with, or simply ignore the realities of nuclear technology. Kristen Potter Farnham writes, "although the damaging effects of past testing cannot be eradicated, this tenacious and persistent activism has helped to create the possibility that the future may be free of these harms."[54] While it is inaccurate to suggest that end of nuclear testing points to a harm-free future—as Solnit shows, nuclear waste poses a perpetual threat—Farnham's gesture towards "possibility" is the touchstone of both Williams' and Solnit's activism. In thinking an unthinkable future, in imagining a world not dominated by apocalypse, both women push for an alternative legacy—one of hope and endurance in the face of bodily, environmental, and social damage. By representing the realities of a nuclear world playing out on a local scale, Solnit and Williams posit an alternative, activist inheritance.

Solnit writes in *Savage Dreams*:

> I have always been plagued myself by an inability to plan for a future that forks so clearly: Up one fork, it makes sense to build a career in anticipation of a comfortable maturity; on the other, I should learn to scavenge and survive in the ruins of a devastated country. This clouded horizon shadows the decisions of most of my peers and unsettles the simplest acts. It was this possibility that made it so easy for many to take a third path: to become activists who pursued no personal comfort but the cancellation of the apocalypse.[55]

The paralysis Solnit describes in terms of the first two forks has plagued the nuclear narrative of both texts. Either one must ignore the future, and carry on as normal, or one must accept the future, and accept with it the impossibility of any kind of "normal." With the third fork, though, Solnit reveals another option: action. Taking this option, Solnit neither denies nor accepts the apocalyptic nuclear future, and instead pursues its cancellation. She is not naïve; as the rest of her text shows, she is not ignorant of the realities of the nuclear future even now—the dust mixed with fallout already covers the Nevada desert, and nuclear waste storage poses problems millennia into the future. Rather, Solnit seeks to halt the progression of the slow apocalypse by calling for an end to any further nuclear activity.

Both texts end with these narratives of activism. Williams recounts her arrest for trespassing as she crosses the cattleguard onto the Nevada Test Site and frames her story—like her recollection of the nuclear flash—in terms of a dream. She writes, "one night, I dreamed women from all over the world circled a blazing fire in the desert. They spoke of change."[56] This change situates her narrative within a hopeful, collective imagining of Solnit's third fork. The collective, or connective, narrative is important to Solnit too, as her text also ends with a circle of people: "the circle became so large its far side was out of sight behind tents and the rolling terrain."[57] At the Columbus Day action at the test site in 1992, which saw impassioned talks by indigenous downwinders, Solnit witnesses in reality the circle Williams dreamed, or imagined, a few years before.

Reading both texts side by side is thus an important part of the legacies with which Solnit and Williams contend. If both women imagine a new future in which nuclear testing is ended and the effects of nuclear waste are, at the very least, no longer replenished by new tests, their writing acts as an extension of the same circles they describe at the actions at the test site. Writing is another, connected, way of thinking a new future, of acting against the dominance of the government's narrative. Williams understands this even at the moment of her arrest:

> As one officer cinched the handcuffs around my wrists, another frisked my body. She found a pen and a pad of paper tucked inside my left boot.
> 'And these?' she asked sternly.
> 'Weapons,' I replied.[58]

As writer-activists, both Solnit and Williams understand the power of their presence at the protests, and their records of them afterwards. They also invoke Ammons' belief that "[texts] play a profound role in the fight for human justice and planetary healing… Words on the page reach more than just our minds. They call up our feelings. They call out to our spirits. They can move us to act."[59] If Solnit and Williams spend the majority of their texts contending with the legacies of physical nuclear fallout in the bodies and on the land of the desert west, and, by extension, the legacies of technoscientific power that stretch back to the beginning of the Manhattan Project, they use the endings of their texts to emphasize the emotive fallout that prompted them to act, and which begins their own legacy. Their texts not only describe action, they provoke it, by calling up our feelings, our outrage. Writing is their way of pursuing a future beyond the apocalyptic promises of nuclear technology, and they engage their reader too in the circles they describe. Their texts, as links in a chain of people, thoughts, and actions, play a vital part in inviting readers to think an alternative to acceptance and denial. By ending their writing on such forward-looking, inclusive terms, *Refuge* and *Savage Dreams* allow their readers to imagine, and thus take part in, the cancellation of the apocalypse.

Notes

1. Bruce Cameron Reed, *The Manhattan Project: A Very Brief Introduction to the Physics of Nuclear Weapons* (Bristol, UK: IOP Publishing, 2017), 7–1.

2. Daniel Cordle, *States of Suspense: The Nuclear Age, Postmodernism, and United States Fiction and Prose* (Manchester, UK: Manchester University Press, 2008), 2.

3. Rob Nixon, *Slow Violence and the Environmentalism of the Poor* (Cambridge, MA: Harvard University Press, 2011), 2.

4. James Berger, *After the End: Representations of Post-Apocalypse* (Minneapolis, MN: University of Minnesota Press, 1999), 5.

5. Charles E. Gannon, "Silo Psychosis: Diagnosing America's Nuclear Anxieties Through Narrative Imagery," in *Imagining Apocalypse: Studies in Cultural Crisis*, ed. David Seed (London: Macmillan, 2000), 112.

6. Jacques Derrida, "No Apocalypse, Not Now (Full Speed Ahead, Seven Missiles, Seven Missives)" trans. by Catherine Porter and Philip Lewis, *Diacritics*, 14.2 (1984), 20–31 (26).

7. Nixon, *Slow Violence*, 3.

8. Rebecca Solnit, *Savage Dreams: A Journey into the Landscape Wars of the American West* (Berkeley: University of California Press, 1999), 5.

9. Solnit also notes that soldiers and animals were exposed to radioactive fallout in experiments deliberately designed to test the ability of military uniforms to withstand exposure to radiation. This adds another layer to the nuclear narrative, in which the government knew, but withheld, the effects of their work on the human body, thus rendering these "tests" not merely bomb tests in the desert, but tests on the site of the body involving both (allegedly) willing and unwitting participants. See *Savage Dreams*, 19.

10. Ibid., 138–9.

11. Joseph Masco, "A Notebook on Desert Modernism: From the Nevada Test Site to Liberace's Two-Hundred-Pound Suit," in *Histories of the Future*, ed. Daniel Rosenberg and Susan Friend Harding, 24.

12. Terry Tempest Williams, *Refuge: An Unnatural History of Family and Place* (New York: Vintage, 1992), 282.

13. Ibid., 281.

14. Ibid., 284.

15. Alan Nadel, *Containment Culture: American Narrative, Postmodernism, and the Atomic Age* (Durham, NC: Duke University Press, 1995), 4.

16. It is worth noting that the Radiation Exposure Compensation Act (1990) (U.S. Public Law 101–426) purported to take responsibility for the effects of nuclear testing previously denied or deflected in government discourse. Covering both the radiation-related illnesses of downwinders, and those contracted by (often indigenous) uranium miners, RECA provides compensation for those affected by nuclear testing in the American west. It also doubles as an apology from the U.S. government to those affected. However, so many restrictions apply to claimants—the age by which they must have been exposed to nuclear materials, the length of residence/exposure, and a proven absence of other possible causes such as smoking—that it is doubtful how much RECA can be seen as fulfilling its own compensatory promises. Moreover, as Laura Harkewicz writes in this volume, "no amount of money could relieve the suffering that came from illness, and what counted as illness was limited to specific radiation-related diseases. There was no compensation for a disease that did not fit the radiation-related disease profile, nor for the disease that came from living in chronic fear of becoming sick sometime in the future." As such, while RECA takes some responsibility for the government's obfuscating narratives of radiation and safety, it does little to account for the broader implications of nuclear testing, nor does it contend with the apocalyptic dread that comes alongside radiation exposure, against which both Solnit and Williams are actively working.

17. Solnit, *Savage Dreams*, 148.

18. Ibid., 150.

19. Ibid.

20. Ibid., 152–3.

21. Robert J. Lifton, *Death in Life: Survivors of Hiroshima* (Chapel Hill, NC: University of North Carolina Press, 1991), 109.

22. Berger, *After the End*, 5.

23. Ibid., 218.

24. Cassandra Kircher, "Review: On Nature Writing in the Nuclear Age," *Fourth Genre: Explorations in Nonfiction*, 15.1 (2013), 197–204 (198).

25. Williams, *Refuge*, 3.

26. Heather Houser, *Ecosickness in Contemporary U.S. Fiction: Environment and Affect* (New York: Columbia University Press, 2014), 3.

27. Williams, *Refuge*, 43.

28. Susan Sontag, *Illness as Metaphor and AIDS and its Metaphors* (London: Penguin, 1991), 5.

29. Joseph Masco, "Bad Weather: On Planetary Crisis," *Social Studies of Science*, 40 (2010), 7–40 (28; 29).

30. Williams, *Refuge*, 44.

31. Solnit, *Savage Dreams*, 4.

32. Williams, *Refuge*, 286.

33. Berger, *After the End*, 14.

34. Ibid., 42.

35. Cordle, *States of Suspense*, 14.

36. Solnit, *Savage Dreams*, 16.

37. Yi-Fu Tuan, *Landscapes of Fear* (Oxford: Basil Blackwell, 1980), 5.

38. Lawrence Buell, *Writing for an Endangered World: Literature, Culture, and Environment in the U.S. and Beyond* (Cambridge, MA and London: Belknap, 2001), 50.

39. Williams, *Refuge*, 97.

40. Ibid., 281.

41. Ibid.

42. John Beck, *Dirty Wars: Landscape, Power, and Waste in Western American Literature* (Lincoln, NE: University of Nebraska Press, 2009), 293.

43. Reed, *The Manhattan Project*, 7–9.

44. Solnit, *Savage Dreams*, 120.

45. Ibid., 78.

46. Ibid.

47. Ibid.

48. Masco, "A Notebook on Desert Modernism," 36.

49. Elizabeth Ammons, *Brave New Words: How Literature Will Save the Planet* (Iowa City, IA: University of Iowa Press, 2010), 170.

50. Solnit, *Savage Dreams*, 83.

51. Nadel, *Containment Culture*, 46.

52. Nixon, *Slow Violence*, 15.

53. Joseph Masco, "'Survival is Your Business': Engineering Ruins and Affect in Nuclear America," *Cultural Anthropology*, 23.2 (2008), 361–398 (363).

54. Kristen Potter Farnham, "Grass Roots Activism: Terry Tempest Williams Offers a Model for Change," *Boston College Third World Law Journal*, 12.2 (1995), 443–456 (453).

55. Solnit, *Savage Dreams*, 376.

56. Williams, *Refuge*, 287.

57. Solnit, *Savage Dreams*, 385.

58. Williams, *Refuge*, 290.

59. Ammons, *Brave New Words*, 172.

CHAPTER EIGHT

"We Can't Relocate the World"

Activists, Doctors, and a Radiation-Exposed Identity

Laura J. Harkewicz

INTRODUCTION

Early one spring morning in 1985, members of the international environmental organization Greenpeace placed the epigram noted in the title on a banner outside the main satellite facility at the Kwajalein Missile Range in the Marshall Islands. Kwajalein was, and is, a military base. Kwajalein Lagoon has been used for decades as the splash-down site for missiles fired from Vandenburg Air Force Base in California. The full message on the banner placed at the Kwajalein satellite facility, which was written in both Marshallese and English, was: "We can't relocate the world, Stop Star Wars."[1]

The Greenpeace visit to Kwajalein was part of the organization's "Pacific Peace Voyage." A highlight of the voyage was the relocation of the people of Rongelap from their contaminated home to Mejato Island, about 100 miles southeast of Rongelap. The Greenpeace vessel, the *Rainbow Warrior*, was chosen for the journey.[2] The people of Rongelap had asked Greenpeace to remove them from their home because they feared they had been betrayed by the U.S. government.

In 1954, the peoples of Rongelap and Utirik had been exposed to radioactive fallout from the Bravo hydrogen bomb test—the largest nuclear device ever tested by the United States. Within days, U.S. officials established the Bravo Medical Program, in response to the needs of over 200 Marshall Islanders who had been exposed to this radiation. The Bravo

153

Program continued for over 40 years with two (often conflicting) goals: medical care for the exposed, and research into the human biological effects of radiation exposure.

The people of Rongelap, whom the U.S. government had evacuated shortly after the Bravo event, were returned to their home atoll in 1957, where their medical surveillance by Bravo Program doctors continued. By the 1970s, lingering scientific uncertainty about radiation effects, and general societal distrust in the objectivity of scientists affiliated with the government, provided an audience for activists who supported—some say created—Marshallese claims of human experimentation at the hands of Bravo doctors.

In the Marshall Islands, exposed Marshallese citizens joined forces with other anticolonial, antinuclear, and health activists who were creating media attention focused on the scientific knowledge generated by the Bravo Program. Activists stressed the need for independent (objective) scientific review of data rather than additional data collection. They argued it was not the data that were unreliable, but asserted that the people involved in its interpretation lacked credibility because the work was done within the national laboratory system. Anti-government, antinuclear activists attempted to create a global radiation-exposed identity based on the collective experience (and potential threat) of radiation exposure. They brought together stories about the experiences of atomic veterans, Nevada Test Site "downwinders," and multiple Marshallese atolls to emphasize the dangers of nuclear weapons. The Marshallese adopted this identity of the radiation-exposed to guarantee their access to medical care and compensation.

Using Adriana Petryna's concept of "biological citizenship," this essay describes how the Marshallese mobilized around a radiation-exposed identity to claim membership in a community of individuals who had suffered similar radiation-induced injuries. Petryna describes biological citizenship as being "based on medical, scientific, and legal criteria that both acknowledge biological injury and compensate for it."[3] Biological citizenship also suggests the value of the body in itself and in its ability to bring people together.[4] Biological citizenship in my usage is not belonging to a nation but rather belonging to a collective identity. The power to lay claim to financial compensation and medical care comes from membership in this community.

In my account, admission to the community of exposed individuals was accomplished through activism and by engaging with activists' definitions of exposed peoples. Stuart Hall describes identity as a "production, which is never complete" and "always constituted within, not outside, representation."[5] Activism was both a representation of identity, and part of the continually changing process of identity formation. Two situations were instrumental to both the Rongelapese request for removal from their island and their formation of a radiation-exposed identity: 1) the release of information from the Northern Marshall Islands Radiological Survey (NMIRS), and 2) the approaching end of independence negotiations between the governments of the Marshall Islands and the United States.

NMIRS CONTAMINATION

In 1978, in response to a lawsuit by the people of Bikini, the U.S. Department of Energy (DOE) conducted an aerial survey of the northern Marshall Islands where nuclear weapons testing had taken place in the period from 1946 to 1958. The next year, DOE representatives went to Rongelap and told the people that the northern islands of their atolls were too radioactive to visit. The people had been using these atolls for food-gathering for over 20 years, since their return to their home in 1957.[6] They also learned that the survey results indicated radiation levels on some of the northern islands of Rongelap as high as those on Bikini atoll, where nuclear weapons had been detonated.[7] The people of Rongelap were outraged by this information. They felt the DOE had betrayed them. With the belief that they had been misled by Bravo Program doctors, the people of Rongelap considered whether they should remain on their atoll.

In the meantime, Brookhaven National Laboratory (BNL) Program personnel and DOE administrators scrambled to find out what the people of Rongelap had been told about the use of the northern atolls, in particular what recommendations had been made about the consumption of *Birgus latro*, the coconut crab. Among Rongelapese foods, the coconut crab was considered to have the highest concentrated levels of radionuclides. Which crabs were "safe" to eat had been a long-standing debate between Bravo doctors and the Marshallese, with often confusing recommendations.[8] DOE personnel thought they had told the people they should not eat the crab from the northern atolls. [Nathan] Tony Greenhouse from the BNL Safety and Environmental Protection Division recalled that "he felt travel by Rongelap people to the northern islands was unlikely because it was an exhausting trip. Therefore, his advice to them was that persons who venture north should refrain from bringing crabs back. However, during an expedition if a person was hungry, then a crab could be eaten occasionally." Greenhouse also said that he had told the people not to use the northern islands for year-round habitation, but that "his notebooks did not contain any references to such restrictions."[9]

What Greenhouse thought he said, what he did say, and how the people interpreted it were all separate issues that could be fought over. By their accounts, the Marshallese had used the northern atolls for food gathering because no one had told them not to do so. They had eaten contaminated coconut crabs. They were convinced the DOE had lied to them and that Rongelap island, like the northern islands of their atoll, might be found contaminated at some time in the future. They wanted to leave. When DOE dismissed their complaints, they looked for help elsewhere. At the same time, they used their outrage to fuel a drive to find others who had suffered as they had. Locating and creating a community of exposed peoples would provide ammunition for their battle against the DOE and would bolster their claims for compensation in the Compact of Free Association, which would certify both Marshallese independence and continued affiliation with the United States

COMPACT NEGOTIATIONS

Meanwhile, negotiations for Marshallese independence from the United States were wrapping up. The Marshall Islands had been part of the strategic Trust Territory of the Pacific since the end of World War II. Activism for independence began in the mid-1970s as the various trust territories struggled toward self-government. The Honolulu-based organization Micronesia Support Committee (MSC) worked for a united, self-sufficient, and nuclear-free Micronesia. MSC members stressed that one consolidated regional government could carry more weight in negotiations with the United States, particularly in terms of placing limits on access to the region for nuclear activities.[10]

Although the experiences of the peoples of Bikini and Enewetak were not directly related to the Bravo Medical Program, their histories were often used by activists as evidence of U.S. government and Bravo Program errors. The atolls of Bikini and Enewetak had been used as nuclear weapons testing sites since the beginning of the Nuclear Weapons Program in the Pacific in 1946. The peoples of these atolls had been relocated before testing began, and the people of Bikini had lived in exile since 1946, when the second nuclear test of Operation Crossroads, the underwater Baker shot, contaminated their island. In 1969, an Atomic Energy Commission (AEC) survey suggested that Bikini was safe for resettlement. Soon after, the first Bikinians returned to their atoll to assist in the resettlement project, which included construction of homes by the U.S. Department of the Interior (DOI). As construction continued, more Bikinians began to return.[11]

In 1972, a thorough radiological survey was conducted at Enewetak Atoll as the first step in a massive cleanup effort aimed at resettlement. The Enewetak survey used more sensitive instruments than those used for the earlier Bikini surveys. Doubts about the quality of the previous measurements obtained motivated Bikinians to request a new survey; a request which Secretary of Interior Rogers C.B. Morton described as having been influenced by "various outside groups, many of whom are critics of the nuclear program."[12] At the time, Morton considered a new survey important to the credibility of the various governmental organizations involved in the resettlement program. Yet no one group had the money for the survey: Years of infighting occurred, eventually causing the Bikinians to file suit against the U.S. government over who would pay for it.[13] While they waited for resolution of their court case, whole-body counts done on those living at Bikini showed that some individuals had 137Cs body burdens in excess of maximum permissible levels.[14] In early 1978, American scientists, believing contamination was due to ingestion of locally-grown food, recommended an imported food program. The program was administered by the Trust Territory of the Pacific Islands, established by the United Nations, but it did not have enough ships to provide regular service, and when the imported food ran out, the Bikinians were forced to eat whatever food was available. Later that year, the Bikinians were sent back to Kili Island, where they had lived during their evacuation. The Bikinians' long-awaited return home, only to be relocated again shortly after, was later described by one activist as the "musical chairs fiasco," an experience that was used in activist publicity to promote doubts about U.S. federal authorities' depictions of "safe" cleanup of contaminated areas.[15]

An activist-authored article published in *The Bulletin of Atomic Scientists* asserted that Bikinians had been misled by the "countless radiological surveys of the island—many of which suggest[ed] the Bikinians were unwitting subjects for radiation tests." Statements made in government documents such as, "Bikini Atoll may be the only global source of data on humans where intake via ingestion is thought to contribute the major fraction of plutonium body burden"—although explained by U.S. government scientists as being made "by technical types anxious to know about the transfer of radioactive elements"—only added to activist beliefs that the Marshallese served as experimental subjects.[16] As the 1986 date of termination of the trusteeship agreement approached, the *Bulletin* article, which framed scientific error as evidence for political irresponsibility, was intended to demonstrate that continued affiliation between the two countries was not in the best interests of the Marshallese. Yet, the newly formed Marshall Islands government favored U.S. strategic access to the islands. Anticolonial activists described the efforts for separate status by some of the Marshall Islands government representatives as their determination to "destroy" a united Micronesia and build their own "mini-empires," which they had attempted to achieve either through "subtle arm twisting" or via the direct support of the majority.[17] Antinuclear and independence activists often operated in opposition to the nascent Marshall Islands government and in particular its first president Amata Kabua.

The Marshall Islands established a constitutional government in 1979 with Kabua as its first president, but negotiations continued with the United States about the specifics of the Compact of Free Association. Some Marshallese supporters stressed that, unlike the majority of islands in the Marshallese chain, the northern islands of Bikini, Enewetak, Rongelap, and Utirik had directly suffered injury and loss from the U.S. Nuclear Testing Program in the Pacific. They argued, therefore, that these islands should negotiate separately with the United States and suggested the Marshall Islands government had no right to Compact compensation. Jonathan Weisgall, the long-time attorney of the people of Bikini, considered it a U.S. government obligation to give Bikinians unique consideration in political negotiations between the Marshallese and U.S. governments. Weisgall took offense at the Marshall Islands government acting as the recipient of any special compensation that might be due the Bikinians. "Interposing the Marshall Islands government in the formulation and administration of a resettlement program," Weisgall argued, "can only cause further bureaucratic snags and squabbles of the kinds that have plagued Bikinians for years." He suggested that the Bikinians favored a resettlement program, perhaps even a form of government, that was under the "direct supervision of the U.S. government." He stated that the Bikinians had "voted for the losers" during the Marshallese elections, and they doubted that those elected had their best interests in mind.[18]

Marshallese President Kabua's family had a long history with the people of Bikini. He was the son of the iroji (chief) whose claim to ownership of Bikini had been rejected by the people years earlier. The Bikinians had not only dismissed the ownership claims of Kabua's father, they had declared their independence from him shortly after they moved away from Bikini in 1946. They considered Lejolan Kabua someone who had only exploited them. Since that point, they had declared that the United States was their paramount chief.[19] In this way, the people of Bikini were using their interactions with the United States to break free from traditional power relations. They saw the interventions of both the newly established Marshall Islands government and right-minded activists who were fighting for a united Marshall Islands as interference, *not* assistance.

Despite continued conflicts among Marshallese supporters, the Marshall Islands government (later the Republic of the Marshall Islands, or RMI) became the sole representative of the Marshallese peoples. While Compact negotiations continued, antinuclear activists used the nuclear histories of all the Marshall Islands, but especially conflicts between Bravo Medical Program doctors and their patient/subjects, as ammunition for their antinuclear activism.

In an article published in the *Los Angeles Times*, journalist Larry Pryor used evidence from an internal Bravo Medical Program report to frame the Marshallese story as an exemplar of medical colonialism and the public health hazards of the nuclear age.[20] The report had stressed the difference in expectations between the people's needs and the stated goals of the program, goals that had been established when the program began in 1954. At that time, when little was known about the human biological effects of radiation exposure, the program had stressed broad-ranging medical care and close observation of all illness. As time went by, and more became known about the effects of radiation on human health, program doctors provided less care, and looked only for the effects they expected. Meanwhile, the Marshallese people had come to understand—from program doctors—that the long-term effects of radiation exposure remained uncertain, suggesting that a long-term general but comprehensive health care program was warranted. This type of medical program was not the mission of the Bravo Medical Program; thus conflicts developed between the two parties.[21] Activists used the Marshallese story as a model for antinuclear/anti-government activism. "It occurred to me quite early," former Peace Corps Volunteer and activist Glenn Alcalay wrote, "that the story of the Marshallese was one which needed to be publicized in terms of the ever-increasing threat of possible nuclear war—a war which will leave no one untouched." Activists' plans were to "reconstruct the history of the nuclear weapons tests in the islands." The result would be to give a "human face" to the problem of "the spread of nuclear weapons in the Cold War period."[22]

In 1983, the people of the Marshall Islands voted on the Compact of Free Association. The Compact allowed the United States to have continued access to the Kwajalein Missile Range. The Compact, although approved by the Marshallese, had yet to be ratified by the U.S. Congress, leaving the possibility for further modifications open. The Section 177 Agreement was the portion of the Compact that dealt with compensation for personal injury and property loss from the Nuclear Weapons Program in the Pacific. Antinuclear activists considered the Section 177 Agreement a "disaster."[23] Not only was compensation inadequate, but activists were also concerned because the Compact removed the islands from United Nations oversight. "As the Compact removes Micronesia from the scrutiny of the international community," they claimed, "Section 177 removes the United States from its role of responsibility for the health and well-being of the victims of its aggressive and militaristic antics."[24] Activists feared, without international oversight, the U.S. government would take advantage of the nascent Marshallese government. Activists' worries that the Marshallese exposures would be invisible once the Compact was ratified prompted them to formulate a community of exposed peoples.

THE COMMUNITY OF EXPOSED PEOPLES

The Bravo Medical Program had provided a dichotomy between the exposed and unexposed that was useful to the Marshallese. The peoples of Rongelap and Utirik received medical care from Bravo Program doctors for radiation-related illnesses and regular sick calls, time permitting, but peoples from the other atolls did not. Being identified as exposed entitled island inhabitants to benefits that unexposed individuals did not receive. As Compact negotiations approached their conclusion, being recognized as exposed became more salient in terms of who would or would not receive compensation. But being identified as exposed might not be enough; public recognition of exposure (publicity) was a better guarantee of benefits. Demands for financial compensation, medical care, and funding for independent studies were based on the vocabulary created by developing a community of exposed peoples, which also served as the touchstone for antinuclear campaigns.

Activist Gifford "Giff" Johnson was central to these campaigns. Johnson had been one of the founding members of MSC. A professional journalist, he edited its journal until the MSC merged with the Pacific Concerns Resource Center in 1983. In 1982, he married Darlene Keju, a Marshallese who was also active in the MSC. Keju, who was born on Ebeye near Kwajalein, claimed she had developed tumors from radiation exposure although she had not been exposed to Bravo fallout. In 1983-84, the Johnsons traveled around the United States and Canada on a speaking tour meant to educate American audiences about Micronesian issues, and to protest against racial prejudice, nuclear weapons testing facilities, and the loss of the political freedom in the Marshall Islands. They also hoped to gain the support of medical professionals for an independent review of the medical program.[25] Keju's tumor-riddled body served as a stand-in for all damaged Marshallese. The Johnsons' presentations not only involved Keju's personal experiences as an Islander exposed to the physical and cultural consequences of nuclear weapons testing, but they also included a slide show featuring graphic photos of radiation burns on the heads, necks, and feet of children and adults exposed to Bravo fallout.[26]

Although the Johnsons' tour provided a physical manifestation of the Marshallese story to the world, in a sense, the community of exposed peoples originated with the 1971 GENSUIKIN (Japan Congress Against A and H Bomb) visit to the Marshall Islands. A medical survey team affiliated with the organization had been invited to the islands by Congress of Micronesia representative Ataji Balos, but Trust Territory administrators had forbidden them access to the outer islands because they were traveling on tourist visas. Although the team had not been allowed to visit the exposed populations of Rongelap and Utirik, they were able to speak to those members of the exposed populations who were living in the capital city of Majuro. The GENSUIKIN visitors considered their trip to have been a great success because their "direct exchange with the victims of the hydrogen bomb test" allowed them to develop "ties of solidarity" with the people of the Marshall Islands. They considered the exposed people to be held hostage by Bravo doctors, and it was the duty of all those who were against nuclear weapons and for world peace to insure their story was told.[27] Charges of human experimentation began with Japanese antinuclear advocates who had made similar claims against the Atomic Bomb Casualty Commission because of their no-treatment policy.[28]

Several years later, in an issue of the newsletter published by JISHU– KOZA (Japan "Stop the Pollution Export" Committee) that critiqued Japan's plans to dump nuclear waste in the Pacific, Giff Johnson reported on the history of nuclear weapons testing in the Marshall Islands. The article was titled, "Marshall Islands: The Radioactive Trust." As a sidebar to his article, Johnson provided a brief account of a U.S. congressional committee that concluded the federal government had "deliberately concealed" the dangers of radioactive fallout exposure from people living around the Nevada Test Site. "The committee criticized the government," wrote Johnson, "'because the agency charged with developing nuclear weapons was more concerned with that goal than with its other mission of protecting the public from injury.'" He concluded his story with the committee's recommendation to compensate "the victims of nuclear testing fairly and without delay."[29]

Johnson's publication of his article in a Japanese newsletter implied his affiliation with the Japanese antinuclear movement. His linking of downwind victim compensation with the history of nuclear testing in the Marshall Islands demonstrated his understanding that their experiences were similar. His conclusion that the congressional committee had recommended compensation did not specify Nevada downwinders, but instead extrapolated that all "victims of the nuclear testing" be compensated "fairly and without delay." If the people in Nevada deserved compensation, so did the Marshallese. Johnson not only drew a correlation between the Marshallese exposure and that of downwinders, but also connected atomic veterans. In correspondence with a veteran who had recently sent him a copy of "Atomic Veterans' Newsletter," he wrote, "The stories of U.S. government treatment of exposed service-men is [sic] very similar to that of the Marshallese. It is a constant battle in which the government continues to say 'prove it' despite the overwhelming evidence." He noted how "useful" it was for the people of the Marshall Islands to learn of the exposure of others.[30]

A common theme in articles about the downwinder exposure and that of atomic veterans was the idea of a government cover-up. Johnson reported that two of the U.S. Air Force weather personnel who had been exposed to Bravo fallout released statements indicating that U.S. officials had decided to go ahead with the Bravo test even though they were aware the winds were blowing in a direction that would contaminate inhabited atolls. The two veterans, Gene Curbow and Donald Baker, stated that the Marshallese had been fortunate to receive medical follow-up because they (the veterans) never had.[31] Carl Schumacher, another veteran of the Bravo test who had served on a decontamination ship during seven separate nuclear tests, reported that he had received a letter from the government five years prior telling him he had been exposed to large amounts of radiation. "We were in a war, but you couldn't see the enemy," reported Schumacher. "What they did was shoot each of us with an invisible bullet."[32] Johnson wrote and collected articles about atomic veterans and downwinders to describe a pattern of U.S. government action (or inaction) toward exposed individuals. This pattern was one of cover-up, minimizing effects, and victimization. Johnson saw and reported connections between these diverse groups to emphasize that the Marshallese were not alone in their suffering. He paid particular attention to what other exposed "citizens" had experienced, and what compensation they had or had not received. He was collecting these experiences to create a profile of exposed peoples.

Activists' activities were not restricted to gathering and writing about information, however. In 1980, John Anjain, the magistrate of Rongelap at the time of the Bravo exposure, and his nephew, Julian Riklon, joined other radiation victims at the National Citizens' Hearings for Radiation Victims held in Washington, D.C., in events that included congressional testimony. In his testimony, Anjain mentioned the NMIRS, and that the survey had shown that the northern atolls of Rongelap were highly contaminated. He claimed, "we have been eating the food and living on these islands since the AEC told us it was safe in 1957." He acknowledged that the DOE had restricted coconut crab ingestion, but, according to Anjain, they had never told the people not to use the land. He wanted to know "why the Congress of the United States is not doing anything for the people of Rongelap," and if it was "possible that the DOE has not fully informed the Congress about their findings."[33] In his brief statement before fellow exposed citizens, Anjain carefully articulated the pattern of government behavior toward exposed peoples that Johnson had documented. Using scientific data, and claiming miscommunication and error, he suggested a cover-up because, surely, if they knew of their problems, the U.S. Congress (and the American people) would not deny the Marshallese the same protections they had extended to other exposed American citizens. Anjain was becoming a radiation-exposed person as defined by the experiences of the community of exposed peoples. Anjain, and other Islanders, mobilized this identity to gain public support for compensation and other assistance, including relocation from their contaminated home. After several years of asking the DOE to move them with no success, the people of Rongelap requested assistance from the international environmental organization Greenpeace. Greenpeace took advantage of the request to promote their campaign against nuclear weapons in the Pacific. By publicizing their antinuclear campaign, which included the relocation of the people of Rongelap, Greenpeace promoted the Marshallese radiation-exposed identity.

THE GREENPEACE RELOCATION AND PACIFIC PEACE VOYAGE

Greenpeace's *Rainbow Warrior* was a 150-foot ketch-rigged ship that had formerly been a fisheries research vessel. The vessel was named for a Cree Indian story in which the rainbow people appeared to lead the earth after its animals and fish had been poisoned. The mission to relocate the people of Rongelap was considered only the "first major campaigning port of call" for Greenpeace's Pacific Peace Voyage. The use of the word "campaigning" in the announcement about the ship's itinerary suggested the trip was imagined as a means to recruit members to the antinuclear cause. Greenpeace's involvement with the relocation would inform the world of the plight of the people of Rongelap—who after 30 years still suffered the damages of radioactive fallout contamination—while serving as "the strongest argument one can make in support of the abolition of nuclear weapons from the face of the earth."[34] The voyage was designed not only to protest U.S. President Ronald Reagan's Strategic Defense Initiative, or SDI, commonly known as "Star Wars," but also to draw attention to French nuclear testing in Polynesia, in hopes of garnering support for the Comprehensive Test Ban Treaty. The final destination of the vessel was to be a three-month vigil off Moruroa, French Polynesia.[35]

When the ship docked in Majuro, Marshall Islands, Senator Jeton Anjain, brother of John Anjain and the Rongelap representative in the Marshallese Nitjela (Parliament), joined the group. In 1984, the Marshallese Nitjela approved the Rongelapese relocation, but they did not provide any funding for the move.[36] In March 1985, in response to Anjain's appeal for assistance, the U.S Congress recommended a $300,000 grant for an independent survey of Rongelap as well as a $3.2 million grant for resettlement to Ebadon. Both grants, however, depended on passage of the Compact of Free Association. Jeton Anjain was worried they would never see these funds and felt his people had waited long enough. When Giff Johnson told him about Greenpeace's plan to visit Kwajalein to protest Star Wars, Jeton asked organizers to help with the evacuation. Greenpeace was "delighted to help and also find a focus for the anti-nuclear voyage."[37] Rongelapese anticolonial activists campaigned with Greenpeace to block the Compact of Free Association, to demonstrate it had been "negotiated in the dark."[38] They believed their activism might still be able to impact the compact before its signature into law.

Like their earlier counterparts, Greenpeace and the Rongelapese protestors used the nuclear histories of Bikini and Enewetak in their publicity. In this case, however, they stressed the histories of the Bravo fallout exposure and its aftermath, a past that featured the Rongelapese experience. The people of Rongelap believed they had the largest claim to the radiation-exposed identity, as their dose had been the highest. Claiming this identity became extremely important in the context of the events of the time. Shortly after the U.S. Congress approved the compact in early 1985, the people of Bikini were awarded a 15-year, $75 million settlement of claims that included an additional $110 million trust fund earmarked for cleanup and resettlement.[39] To a large extent, the award was due to attorney Jonathan Weisgall's relentless pursuit of legal claims.[40] The compact contained a provision that it was to serve as the settlement for "all claims, past, present and future" between the two countries or its citizens in relation to the Nuclear Testing Program, so once the compact was signed into law, no additional legal claims were justifiable.[41] In light of these conditions, the people of Rongelap knew they had to act immediately to garner public support for their cause. But the Bikini award was more personal than just its legal and political implications in terms of the compact. As John Anjain poignantly observed, it was the people of Rongelap, not the people of Bikini, who had "actually suffered from radiation" yet they were "shunned" by the United States government. Many Islanders could not "understand why the U.S. government ha[d] apparently turned its back on them," leaving them with no choice but to abandon their home.[42]

The *Rainbow Warrior* arrived on Rongelap on May 18, 1985. Two days later, during his last service to a packed audience at the local church, Pastor Jatai Mongkeyea compared the Greenpeace evacuation to "the Biblical deliverance of the Israelites from Egypt." Greenpeace members described themselves as "reluctant, mystical saviours emerging from across the horizon to pluck a community from doom."[43] The relocation, dubbed "Operation Exodus," took several days. The land the Rongelapese relocated to was provided by the family of Amata Kabua, the president of the Marshall Islands.[44] The Kabua family was the iroji, or chiefs, of Rongelap. As noted earlier, the people of Bikini had rejected the Kabua family claim to their atoll; this may be one reason why they appeared to be faring better in terms of compensation than the Rongelapese—they had rejected the traditional power structure in favor of another. One attorney who had represented both the Marshallese and Americans (at different times), said "Bikini is the only community out there that has parlayed its experience into what resembles an alternative form of self-sufficiency."[45] They had done this, not only through rejection of the traditional power structure, but also through intensive engagement with the U.S. legal system and putting their contamination experiences to use. "There's a dark joke in Washington," said the unnamed attorney, "that Bravo was the best thing that ever happened to the Bikinians."[46] Prior to the Bravo test, nuclear tests had been moved from Bikini to Enewetak, allowing the radionuclides from the 1946 Operation Crossroads test to decay. The hydrogen bomb tests, which had much larger yields, were moved back to Bikini because the military did not want to threaten the military facilities located at Enewetak. The contamination from the Bravo test made Bikini uninhabitable once more, a condition Bikinians eventually used to their favor.

The people of Rongelap, however, did not fare as well. They were still dependent on the atoll's traditional power structure. As the Rongelap iroji, it was Kabua's responsibility to care for the people. Why would the Marshallese approve relocation, but not provide funds for relocation? Kabua and the Marshallese government wanted to play both sides of the fence. They would allow the people to move, even give them the land to relocate, but they would make relocation difficult because it was the responsibility of the U.S. government, who had left the people contaminated, to pay for their relocation. In the meantime, they could try to delay the relocation until the compact passed. By giving the people of Rongelap land for relocation, Kabua served his iroji role, but his interests were best served by passage of the compact. As iroji, the Kabua family received one-third of any compensation the people of Rongelap were awarded. This meant that, once the compact was official, 33 percent of the $2.5 million scheduled to go to the people of Rongelap would go to the Kabua family.[47]

Jeton Anjain knew this. He believed getting Greenpeace involved in their relocation would mean that the massive amount of publicity generated against Star Wars and the compact might also serve Rongelapese efforts to alert the people of the world that it was the people of Rongelap who suffered the most from the U.S. Nuclear Testing Program in the Pacific. The people of Rongelap sided with Greenpeace to stake their claim to a radiation-exposed identity. John Anjain, who had become the unofficial spokesperson for the people of Rongelap, gave a voice to this identity. At 63 years old, after traveling across the world as a witness of his people's victimization, and having lost his son to leukemia, Anjain had seen it all. "I know the scientists will not come out and tell the truth," he said. "I know from my own experience that we have a serious radiation problem in Rongelap."[48] These were the people whose lives had been shaped by their radiation exposure. They were the radiation-exposed and they based their identities in opposition to U.S. government science and scientists, a definition that antinuclear protestors like the rainbow warriors embraced as well.

While Greenpeace continued their antinuclear and environmental campaigns on a variety of fronts, Marshallese activism became more diffuse once the compact became law in January 1986, when President Ronald Reagan signed the agreement. In October of that year, Public Law 99-239 became effective, granting the Marshall Islands independence from the United States, and officially establishing the Republic of the Marshall Islands. The Section 177 Agreement of the compact awarded the peoples of Bikini, Enewetak, Rongelap, and Utirik $75 million, $48.75 million, $37.5 million, and $22.5 million, respectively. The awards were to be paid in quarterly installments over a 15-year period, after which negotiation between the two countries could be reopened. The compact also made available $3 million to conduct medical surveillance and radiological monitoring. It established the Nuclear Claims Tribunal (NCT), an independent judiciary charged with determining who was eligible for compensation as well as the appropriate monetary distribution. At the time the compact took effect, the tribunal had very little specific information about individual doses and the extent of contamination. They therefore made two presumptions: first, that residence in the Marshall Islands at the time of the Nuclear Testing Program was sufficient evidence of exposure levels that might induce a medical condition, and, second, that any medical condition listed on the compensation schedule was presumed to have been caused from radiation exposure.

Two American programs served as models for the NCT program: the Radiation-Exposed Veterans Compensation Act of 1988 (RVCA) (U.S. Public Law 100-321) and the Radiation Exposure Compensation Act (RECA), commonly referred to as the "Downwinders'" Act (U.S. Public Law 101-426). In these programs, compensation was determined not on a case-by-case basis, but rather by a set schedule of medical conditions or property damage. Using these programs as templates, the NCT developed a six-level classification scheme, with compensation ranging from $12,500 to $125,000. A major part of each award was designated for pain and suffering. The individual radiation dose received was irrelevant to compensation. The Marshallese were not eligible for compensation under these legislations because they were not American citizens. Although the RVCA and RECA mark a difference between the consequences of the Marshallese exposure and that of American citizens, these legislations also mark one way the Marshallese became defined as "pseudo-American citizens." As noted, the Marshallese NCT had used these legislations as models in their development of a Marshallese compensation plan, implying the independent judiciary's intention to compensate the Marshallese *as if* they were American citizens. Marshallese exposures, in fact, had already been considered equivalent to that of American citizens. "The greatest irony of our atmospheric nuclear testing program," noted a 1980 congressional report, "is that the only victims of U.S. nuclear arms since World War II have been our own people. (The Subcommittee includes within this context those individuals living in the trust territories of the Pacific Islands.)"[49]

Despite compensatory awards, most of the exposed Marshallese remained dissatisfied. No amount of money could relieve the suffering that came from illness, and what counted as illness was limited to specific radiation-related diseases. There was no compensation for a disease that did not fit the radiation-related disease profile, nor for the disease that came from living in chronic fear of becoming sick sometime in the future—what Congress of Micronesia representative Ataji Balos referred to as "the ghost of the bomb."[50] Anthropologist Joseph Masco describes this experience as the "nuclear uncanny"—the "inability to evaluate risk in everyday life" that is the result of "proliferating psychic anxiety" due to the "temporal ellipsis" between radiation exposure and disease development.

The inability to separate oneself from one's exposure leads to paranoia, and a form of radioactive manifest destiny in which the development of illness from radiation is inevitable.[51] Therefore, compensation, even if it were deemed appropriate by radiation-related disease designation, did not have the ability to resolve the controversy surrounding exposures to fallout, be they American or Marshallese. The struggle for compensation, however, became another component of the radiation-exposed identity. Identity-formation is based on interactions; it is a process that is always evolving. Therefore, while publicity and activism affected this identity, so did interactions with Bravo doctor/researchers and the government of the United States. Bravo doctors failed to learn the language of the Marshallese. They never translated their reports into Marshallese.[52] The coconut crab confusion demonstrated how researchers failed to both understand the Marshallese diet, and explain the hazards involved with consuming some foods or visiting some areas.

DOE researchers often conducted studies without consulting the people involved for vital information. Bravo doctors claimed the people just didn't understand, yet they never took the time to explain. They dismissed complaints they did not expect as psychological. In his memoir about his experiences as director of the medical program, Robert Conard wrote, "Due to language difficulties, it was not easy to be sure about the interpretation of our remarks; indeed, as time went on, it became apparent that our explanations were not fully understood. This is not surprising. People with more sophisticated backgrounds than the Marshallese have difficulty understanding radiation and its effects, and as with the Marshallese, the issue is often charged with emotion."[53] Was it the Marshallese who did not understand or the doctors? Or both? Whatever the rationale for their behavior, it is clear that researchers failed to comprehend how their actions impacted the people, often making interactions between the two groups worse. It was not a simple matter of misunderstanding, however. Marshallese reactions, and the identity(s) produced thereafter, were complicated by multiple layers of interactions with many people over many years.

Although this analysis of Marshallese activism and identity formation is unique, it has similarities to that described by scholars of other advocacy groups. Adriana Petryna describes how those injured by the Chernobyl accident used knowledge of biological damage as "a means of negotiating public accountability, political power, and further state protections in the form of financial compensation and medical care."[54] Similarly, Phil Brown and his collaborators discuss how community-based environmental justice organizations created a "collective identity around the experience of asthma," which they linked to social disadvantage in order to promote social reform.[55] Like the victims of Chernobyl or various patients' rights groups, the Marshallese attempted to create an identity—that of the "radiation exposed"—that they tried to use to their advantage. Meanwhile, Bravo Program doctors implied, "If we cannot understand your disease by the tools of our discipline, then you must be at fault."[56] Bravo scientists didn't see their part in adding to the people's suffering. The result was the creation of an identity constituted by and through these experiences.[57] This identity was based not only on experience, but also representations of activism. Anticolonial, antinuclear activists stressed the politics of the science of the Bravo Medical Program—the conflict of interest and lack of objectivity inherent in the program's dual goals. Activists' representations stressed the research goal of the Bravo Medical Program and identification with experimental subjects. The medical goal was lost.

The radiation-exposed identity and the publicity generated by activism gave sufferers the opportunity to challenge national laboratory scientists' "monopolistic claim over knowledge."[58] Unfortunately, the scientific knowledge about illnesses caused by radiation exposure was, and remains, indeterminate, making activists' use of science "both a curse and a point of leverage."[59] Uncertainty left radiation effects open to interpretation, while legal and political claims were based on scientific and medical definitions that stressed certainty. After the Compact of Free Association was signed into law, science, not politics, took the lead. Various scientific studies, many conducted by independent scientists, took place in the Marshall Islands. The majority of these were related to reassessment of contamination, cleanup, and resettlement. Science, whatever its shortcomings, offers a greater promise of objectivity than politics. As their contaminated islands were made ready for habitation, the people still wondered if their homes could ever be truly safe. Objectivity, whatever its promise, is not the same as certainty.

Notes

1. Greenpeace Scrapbook, Pacific Manuscript Bureau, PMB 1238, Research School of Pacific and Asian Studies at the Australian National University, accessed via member library, University of California, San Diego Geisel Library (hereafter referenced as GPSB) photo 2, 37.

2. The Marshall Islands are located about midway between Hawaii and Japan in the Pacific Ocean. The atolls of Enewetak and Bikini were used for nuclear tests from 1946 to 1958. Rongelap is about 50 miles south of Bikini. Enewetak is approximately 200 miles west of Rongelap and Bikini. Utirik is located about 250 miles east of Rongelap and Bikini. Kwajalein, where a U.S. military base is sited, is located approximately 200 miles southeast of Rongelap and Bikini. Majuro atoll and the city of Majuro, where the majority of Marshallese live and where the Republic of the Marshall Islands government is located, is about 250 miles southeast of Kwajalein.

3. Adriana Petryna, *Life Exposed: Biological Citizens after Chernobyl* (Princeton, NJ: Princeton University Press, 2002), 6.

4. Nikolas Rose and Carlos Novas, "Biological Citizenship," in *Global Assemblages*, ed. Aihwa Ong and Stephen J. Collier (Oxford, UK: Blackwell Publishing Ltd., 2005) 439–441. Also see: Catherine Waldby, *The Visible Human Project: Informatics, Bodies, and Posthuman Medicine* (London, UK: Routledge, 2000).

5. Stuart Hall, "Cultural Identity and Diaspora," in *Identity: Community, Culture, and Difference*, ed. Jonathan Rutherford (London, UK: Lawrence and Wishart, 1990), 222–237, 222.

6. Giff Johnson, "Paradise Lost." *The Bulletin of the Atomic Scientists* (December 1980), 24–29, 28.

7. Letter from Giff Johnson to Senator Edward Kennedy, 12 July 1979, Reel 1, Gifford Johnson Papers, Pacific Manuscript Bureau (PMB) 1172 (hereafter referenced as "GJP"); Letter from Giff Johnson to Julian Riklon, Ebeye, Marshall Islands, 24 August 1979, Reel 1, GJP.

8. Transcript for the "Symposium for American Association for the Advancement of Science (AAAS)," 7 January 1980, Reel 4, GJP.

9. Letter from Edward T. Lessard, Marshall Islands Radiological Safety Program Director to Tommy McGraw, U.S. DOE, Division of Health and Environmental Research, 24 July 1981, Document # 402930, Department of Energy Office of Health Programs Marshall Islands Document Collection, http://worf.eh.doe.gov (hereafter referenced as "OHP").

10. Robert C. Kiste, "A Critical Eye on Uncle Sam." *Pacific Islands Monthly*, 1984, Reel 2, GJP.

11. Jonathan M. Weisgall, "The Nuclear Nomads," *Foreign Policy* 39 (Summer 1980): 74-98.

12. Weisgall, "The Nuclear Nomads," 87.

13. Weisgall, "The Nuclear Nomads."

14. Cesium-137 is one of the more common fallout products created by nuclear fission of Uranium-235 in nuclear weapons.

15. Letter from Glenn Alcalay to Senator Adlai Stevenson, Jr., 10 July 1980, Reel 1, GJP.

16. Giff Johnson, "Micronesia: America's 'Strategic' Trust." *The Bulletin of the Atomic Scientists* (February 1979): 11–15, 15.

17. Letter from Hans Wiliander to Giff Johnson, 2 August 1977, Reel 2, GJP; Letter from School of Area Studies, The University of Tsukuba, Japan to Giff Johnson, 1 September 1977, Reel 1, GJP.

18. Weisgall, "Nuclear Nomads," 96.

19. Jack A. Tobin, "Appendix I: Preliminary Anthropologist's Report–Bikini Atoll Survey 1967," Document # 408366, OHP.

20. Larry Pryor, "U.S. Forced to Restudy Radiation Peril: Islanders stricken 23 years after H-test," *Los Angeles Times*, 11 June 1977.

21. Konrad Kotrady, "The Brookhaven Medical Program to Detect Radiation Effects in Marshallese People: A Comparison of the People's vs. The Program's Attitudes." 1 January 1977. Document # 403773, OHP.

22. Glenn Alcalay, "The Tropics of Cancer: Concept Page," 12 July 1982, Reel 6, GJP. Alcalay served with the Peace Corps on Utirik from 1975 to 1977.

23. Letter from Giff Johnson to Walden Bello and Glenn Alcalay with "Some Comments on the Washington, D.C. Network Proposal" attached, 23 August 1984, Reel 2, GJP.

24. U.S. Pacific Issues Network Confidential Memorandum to Nuclear-free and Independent Pacific (NFIP) Executive Committee and National Representatives, from Walden Bello and Glenn Alcalay, re: NFIP's Position on the Compact of Free Association, 4 August 1984, Reel 2, GJP.

25. Jack Swanson, "Islander Arrives to Protest U.S. Nuclear Missile Tests," *Seattle Post-Intelligencer*, 2 January 1983, Reel 2, GJP; Nadine W. Scott, "Tumor Patient Claims U.S. Cover-up," *Honolulu Star Bulletin*, 3 September 1983, Reel 4, GJP; "Report on the United States/Canada Speaking Tour of Giff Johnson and Darlene Keju, February to April 1984," Reel 2, GJP; Barbara Walters, "Islands Tired of 'Guinea Pig' Role," *Kalamazoo Gazette*, 16 February 1984, Reel 4, GJP.

26. Creg Darby, "What a Drag it is Getting Bombed: Islanders Claim Suffering Due To 50's A-Bomb Tests," *West Washington Daily*, 12 January 1983, Reel 4, GJP

27. The GENSUIKIN Survey Team to Micronesia from Dec. 7 to 18, 1971, "Report on the Investigation of Damage Done by the Bikini Hydrogen Bomb Test to the People of the Marshall Islands," GENSUIKIN NEWS, 15 February 1973, 13, Reel 4, GJP.

28. Giff Johnson, "Downwind Danger," in JISHU-KOZA/Japan "Stop the Pollution Export" Committee, "Don't Make the Pacific a Nuclear Dumping Ground!, Part 2," 2 September 1981, Reel 2, GJP.

29. The bombings of Hiroshima and Nagasaki were considered an act of war. The U.S. government considered treatment of the injured an admission of guilt, therefore they only studied the exposed, leaving medical care to Japanese physicians. For more information on the ABCC's no-treatment policy see: Susan Lindee, *Suffering Made Real: American Science and the Survivors of Hiroshima* (Chicago: The University of Chicago Press, 1994), 117–142.

30. Letter from Giff Johnson to Thomas Saffer, 3 March 1981, Reel 1, GJP.

31. Giff Johnson, "So What's New: Another Nuclear Cover-up," *Micronesia Support Committee Bulletin* (Summer 1981), Reel 4, GJP.

32. John Christensen, "Like the Gates of Hell," *Honolulu Star Bulletin*, 10 May 1984, Reel 4, GJP

33. John Anjain, Testimony to the U.S. Senate Committee on Energy and Natural Resources as quoted in George R. Blake, "It was a bomb… I know now…saw him die," *Honolulu Star Bulletin*, 16 June 1977.

34. GPSB, 3–4.

35. GPSB, 6, 7.

36. David Robie, "Nuclear deliverance," *Islands Business*, May 1985 (GPSB 28).

37. GPSB, 27, 34, 13.

38. David Robie, "The Legacy of Bravo Monster," *New Zealand Times*, 5 May 1985 (GPSB 21).

39. Jonathan M. Weisgall, *Operation Crossroads: The Atomic Tests on Bikini Atoll* (Annapolis, MD: Naval Institute Press, 1999), 315.

40. Jeffrey Davis, "Bombing Bikini Again, (This Time with Money)," *New York Times Magazine*, 1 May 1994, 69.

41. Republic of the Marshall Islands, "Petition Presented to the Congress of the United States of America Regarding Changed Circumstances Arising from U.S. Nuclear Testing in the Marshall Islands," 11 September 2000, Attachment III: Legal Analysis, 3, OHP.

42. Robie, "Nuclear deliverance," (GPSB 28).

43. GPSB, 42.

44. GPSB 32; Jane Dibblin, *Day of Two Suns: U.S. Nuclear Testing and the Pacific Islanders* (New York: New Amsterdam, 1988), 185.

45. Davis, "Bombing Bikini Again,"69.

46. Ibid.

47. Diblin, *Day of Two Suns*, 185.

48. GPSB, 30–31.

49. United States Congress Subcommittee on Oversight and Investigation, "The Forgotten Guinea Pigs," August 1980, 37.

50. Ataji Balos, Congress of Micronesia representative of the peoples of Rongelap and Utirik, as quoted in Roger W. Gale, "Our Radioactive Wards, No One Warned the Micronesians," paper presented at the Conference for a Nuclear Free Pacific held in Suva, Fiji, April 1–6, 1975, Reel 3, GJP.

51. Joseph Masco, *Nuclear Badlands: The Manhattan Project in Post-Cold War New Mexico* (Princeton, NJ: Princeton University Press, 2006), 32.

52. Robert A. Conard, "Summary of the 20th Post Exposure Medical Survey of the Rongelap and Utirik People," Document #402831, OHP.

53. Robert A Conard, "Fallout: The Experiences of a Medical Team in the Care of a Marshallese Population Accidentally Exposed to Fallout Radiation." BNL 46444 Informal Report (Springfield, VA: National Technical Information Service, 1992), 22.

54. Petryna, *Life Exposed*, 7.

55. Phil Brown, Brian Mayer, Stephen Zavestoski, T. Luebke, Joshua Mandelbaum, and Sabrina McCormick, "The health politics of asthma: environmental justice and collective illness experience in the United States." *Social Science & Medicine* 57, no. 3 (August 2003): 453–464, 5.

56. Veena Das, "Suffering, Legitimacy and Healing: The Bhopal Case" in *Critical Events: An Anthropological Perspective on Contemporary India* (Delhi, India: Oxford University Press, 1995).

57. Petryna, *Life Exposed*, 12.

58. Das, "Suffering, Legitimacy, and Healing," 168.

59. Petryna, *Life Exposed*, 28.

CHAPTER NINE

Hanford Production Reactor Operations and Contamination in the Columbia River

M. S. Gerber

INTRODUCTION

The largest and most significant battlefield in America's effort to win World War II and fight the Cold War was the Hanford Site in Washington State. Known first as the Hanford Engineer Works (HEW), and then variously as the Hanford Works (HW), the Hanford Nuclear Reservation, and finally the Hanford Site under different federal agencies, the sprawling tract is both beautiful and damaged. It was chosen by its Army builders for its isolation, mild climate, abundant gravel for concrete, but mostly for its chief asset—the powerful, swift and massive Columbia River. The river flows for 51 miles through and past Hanford, sometimes appearing as a deep indigo, sometimes sparkling cobalt, sometimes a brooding slate gray. It was the crucial resource that cooled the reactors that irradiated uranium to transmute small fractions into plutonium, the business core of the first and most abundant nuclear weapons. While serving its country and cooling the inferno inside the blocky reactors, the magnificent Columbia suffered indignities rarely visited on a treasure so pristine. During the years of reactor operations, many radiological contaminants entered the Columbia River in reactor effluent, affecting aquatic life and the entire food chain. Chemical contaminants and thermal load also entered the river, affecting ecosystems. Hanford scientists attempted to control contaminants with various strategies, but essentially all techniques failed until reactor operations decreased substantially.

Nine plutonium production reactors were built along the Columbia River at the Hanford Site between 1944 and 1963. All were graphite moderated and "light-water" cooled with river water. The first eight Hanford reactors had single-pass (once-through) cooling systems, meaning that water from the Columbia was pumped through their process tubes to cool the fission reaction just once. After racing through the process tubes, the water was routed through large "downcomer" pipes to outdoor retention basins to partially cool and settle before being pumped back into the Columbia River.[1] The ninth production reactor at Hanford, N Reactor, had a recirculating primary coolant system, meaning that water from the Columbia River was pumped through the process tubes, recirculated, and pumped through the tubes repeatedly in a continuous loop. A "feed and bleed" system operated in the recirculation piping to add small amounts of fresh water and "siphon off" small amounts of used water, to continually refresh the system. N Reactor also had four smaller, secondary cooling systems (the shield, control rod, graphite, and fuel storage basin cooling systems) that operated as single-pass systems.[2] All of these reactors taxed the river's vitality to greater and lesser degrees, but none were gentle on the great watercourse.

Fortunately, time has done the work that Hanford scientists could not. Since the production reactors shut down in phased sequence, the radionuclides in the river have decayed to nearly undetectable levels, the thermal load is gone, and a significant burden of the chemical contaminants have washed through muds and eddies and dispersed in the vast Pacific Ocean. The Columbia has reinvented itself and reigns again as the crown jewel of the Pacific Northwest.

Reactor Operating Dates, Water Usage Power Levels

The dates of construction and shutdown of the Hanford production reactors are shown in Table 9.1.

During the mid-1950s through the early 1960s, major upgrades were made at the first eight Hanford reactors, raising the throughput of coolant water from 30,000 gallons per minute (gpm) to 71,000–80,000 gpm at the six oldest reactors and to 115,000 gpm at each of the K Reactors. Total water usage by the Hanford production reactors rose from approximately 90,000 gpm in 1945 to 665,000 gpm by 1964.[3] During the same timeframe, more reactors were being built at Hanford. Higher power levels together with additional reactors compounded the problems of water discharge into the Columbia River. Effluent was hotter, and more laden with chemicals, purge debris from film removals, and radionuclides.

Table 9.1. Dates of Operation, Hanford Production Reactors

Reactor	Start-Up Date	Deactivation Date
B Reactor*	September 26, 1944	February 12, 1968
D Reactor	December 17, 1944	June 26, 1967
F Reactor	February 25, 1945	June 25, 1965
H Reactor	October 29, 1949	April 12, 1965
DR Reactor	October 3, 1950	December 30, 1964
C Reactor	November 18, 1952	April 25, 1969
KW Reactor**	January 4, 1955	February 1, 1970
KE Reactor	April 17, 1955	January 29, 1971
N Reactor†	December 1963	January 1987

* B Reactor was shut down from March 16, 1946 to July 1, 1948
** KW Reactor was shut down from January 5, 1955 to March 11, 1955
† N Reactor was shut down briefly during 1971

Approximate Columbia River water usage by the Hanford production reactors is shown in Table 9.2. Total aggregate Columbia River water usage by Hanford production reactors over time is shown in Table 9.3.

Table 9.2. Columbia River Water Throughput of Hanford Production Reactors (Numbers are approximate)

Reactor	Water Use at Start-up	Water Use by 1964	Total per Day by 1964
B Reactor	30,000 gpm	71,000 gpm	1.02 billion gallons
D Reactor	30,000 gpm	71,000 gpm	1.02 billion gallons
F Reactor	30,000 gpm	71,000 gpm	1.02 billion gallons
DR Reactor	30,000 gpm	71,000 gpm	1.02 billion gallons
H Reactor	40,000 gpm	71,000 gpm	1.02 billion gallons
C Reactor	50,000 gpm	80,000 gpm	1.15 billion gallons
KW Reactor	70,000 gpm	115,000 gpm	1.65 billion gallons
KE Reactor	70,000 gpm	115,000 gpm	1.65 billion gallons
N Reactor*	1,500 gpm	1,500 gpm by early 1970s	2.16 million gallons

* N Reactor's primary recirculating system caused much lower water throughput. Conversion of secondary systems to recirculating cooling systems in 1970s kept water usage relatively low.

Table 9.3. Total Columbia Water Throughput of Hanford Production
Reactors by Hour and 24-hour Day
(Numbers are approximate)

Year	Water Usage by All Reactors per Hour	Water Usage, All, per Day
1945	5.5 million gallons	1.29 billion gallons
1955	21 million gallons	5 billion gallons
1964	39.9 million gallons	9.76 billion gallons

The original three Hanford reactors (B, D, and F), as well as DR Reactor, were designed to operate at 250 megawatts (MW) thermal. H Reactor was designed to operate at 400 MW, C Reactor at 650 MW, and the KW and KE Reactors at 1,800 MW each. N Reactor was designed to operate at 4,000 MW.[4] Increasing the power levels of the production reactors was always a goal of Hanford operators because it enabled increased plutonium production. By the early 1960s, power levels were greatly enhanced.[5] Higher power levels were achieved by slightly enriching the uranium fuel, operating at higher water temperatures and flows, changing the physical configuration of the fuel elements and the construction of the process tubes, installing instrument upgrades and modifications, and altering other operating parameters.[6]

Power levels at the Hanford production reactors are shown in Table 9.4.

Table 9.4. Operating Power Levels of Hanford Production Reactors[7]

Reactor	Start-up Power Level & Year	Power Level by 1964	Factor Increase
B Reactor	250 MW in 1944	2,210 MW	8.8X
D Reactor	250 MW in 1944	2,165 MW	8.6X
F Reactor	250 MW in 1945	2,040 MW	8.1X
H Reactor	400 MW in 1949	2,015 MW	5.3X
DR Reactor	250 MW in 1950	2,140 MW	8X
C Reactor	650 MW in 1952	2,500 MW	3.8X
KW Reactor	1,800 MW in 1955	4,400 MW	2.4X
KE Reactor	1,800 MW in 1955	4,400 MW	2.4X
N Reactor	2,400 MW in 1963	4,000 MW by 1970	1.6X

Note: Pressure levels and operating parameters were different at N Reactor

Operation of Effluent Retention Basins at Hanford Production Reactors

The retention basins of the B, D, and F Reactors were made of concrete, with two halves, each half having a capacity of 6 million gallons. The retention basins of the H and DR Reactors also were made of concrete, with each half having a capacity of 9 million gallons. The C Reactor retention basins were two separate steel enclosures with capacity of 10 million gallons each, and the K Reactors each had three steel retention basins with capacities of 9 million gallons each.[8] All retention basins were outdoors, and open to the air. Effluent water was discharged from the basins via large pipes (called "river lines") that were buried several feet below the riverbed.[9] Disposal practices changed over time, as effluent flows increased markedly and retention basins cracked due to thermal shock and other wear and tear. In early 1947, effluent retention time was listed as six hours.[10] By 1960, two secret[11] reports estimated holdup time from "about 30 minutes to about 3 hours."[12] By the time operations shut down, every one of the effluent retention basins had failed due to leakage, overflow, or both.

Radionuclides of Concern in Hanford Production Reactor Effluent

Inside the process tubes of the Hanford reactors, cooling water became "activated" with radionuclides from the fission process occurring in the uranium fuel in the tubes. Activation occurred when trace elements in the Columbia River water such as iron, chrome, and zinc were bombarded with neutrons and formed radioisotopes (or unbalanced forms) of themselves. These activation products were carried in the cooling water as it passed out of the reactors, into the holding basins, and back into the Columbia River. Of the more than 60 radionuclides in the effluent water of Hanford's production reactors, five were of primary concern. These were phosphorous 32 (P^{32}), arsenic 76 (As^{76}), zinc 65 (Zn^{65}), chromium 51 (Cr^{51}), and neptunium 239 (Nr^{239}). Primarily beta emitters (except Np^{237}, an alpha emitter), these radioisotopes affect the gastrointestinal tract, bones, and reproductive and blood-forming organs. Concern about them arises due to their physical effects, the massive quantities released, and the fact that their half-lives are two weeks (in the case of P^{32}) and longer, meaning that they did not decay substantially in the effluent basins before being discharged back into the Columbia River.[13]

EARLY FISH STUDIES (RADIOLOGICAL) AT HANFORD WORKS

In 1945, Hanford's federal contractor DuPont Corporation began studies of the responses of fish to the radionuclides and chemicals discharged into the Columbia River in reactor effluent, and to water temperature increases from effluent. Biologist Richard Foster established a fish laboratory in the 100-F Area of the Hanford Engineer Works (HEW—World War II name for the Hanford Site), near the river shore.[14] Almost immediately, fish held in troughs containing varying concentration of reactor effluent water showed a "high rate of mortality." The following year, Foster's group discovered several basic trends in the responses of fish to various concentrations of reactor effluent.[15] Importantly, they noted bioaccumulation. In other words, radioactivity accumulated in fish bodies at much higher levels than existed in the river water itself. It also concentrated in fish tissues in varying degrees, being highest in livers, kidneys, and gills, and relatively lower in muscle tissue (flesh). The higher metabolic rates of juvenile trout and salmon caused them to accumulate proportionately more radioactivity in their tissues than did adult fish.[16] Contamination levels were highest from August through November, due to low river flows. All this knowledge was new, groundbreaking, and exciting to the scientists performing the studies, but government policy prevented them from sharing it with the larger scientific community.

In September 1946, the General Electric (GE) Hanford Company took over from the DuPont Corporation as the operating contractor at HEW. As of January 1, 1947, the new civilian Atomic Energy Commission (AEC) assumed management of U.S. nuclear production sites from the Army. At that time, the name HEW was changed to Hanford Works (HW). The fish laboratory became part of the Health Instruments (HI) Division of HW, managed by Herbert Parker.

By spring 1947, Hanford scientists were learning significantly more about fish damage from reactor effluent. Fish laboratory biologist K. Herde found that the concentration factor of fish tissue over that of river water reached as high as 170,000 times and averaged 100,000 times during 1946. "The concentration factor is higher by a factor of several hundred than factors previously reported," he stated. "All tissues showed a pronounced increase in activity." High levels of radioactivity were found in fish taken as far as 20 miles below F Reactor, the reactor furthest downstream. Fish taken from the Yakima River, and from the Columbia River upstream from the Hanford reactors, contained "miniscule to nonexistent radioactivity levels."[17]

In the second half of 1947, the government ordered a major expansion in plutonium production at HW. GE Hanford Co. started construction on the H and DR Reactors, as well as many other facilities. Engineers began increasing the power levels at the existing reactors. Herde reported in autumn 1948 that "river contamination...justifies considerable concern...[There is] a very narrow margin of safety between actual radiation detected and tolerance values...It seems reasonable to assume that some fish tissues occasionally approached the tolerance value of 100 mrep [milli-rep] per day."[18] The most prevalent and highly concentrated radioisotope in fish was P^{32}.[19] That autumn, another HI scientist reported "the concentrating ability of the bottom algae appears to be, in essence, the foundation of the radiobiological problem in the river," and that concentrations of radioactivity in algae and plankton were 1,000 to 4,000 times that of the river water.[20] Another study reported that "all aquatic organisms collected below the pile [reactor] areas which have been analyzed have contained radioactivity many times that found in the surrounding water of their environment."[21] A concurrent study of pekin ducks from the Columbia River near F Reactor showed that P^{32} was the "dominant isotope present" in tissues at sacrifice.[22] Water samples from the Columbia River at HW taken during the last quarter of 1948 demonstrated radioactivity levels "higher by factors of 2 to 3 at most locations in the river as compared with the results of the previous quarter."[23] The problems were growing.

During the following years, as more reactors started up and power levels rose, HW scientists observed increased radioactive contamination in both the Columbia River water and its organisms. The brevity of this essay does not allow for full exposition of various reports, most of which were secret. However, a few sample reports will be quoted to give a measure of the steep climb in river contamination. In 1949, Foster explained that:

radioactive materials which have been accumulated in the [river] plants are consumed by herbivorous forms including small crustaceans, insect larvae, and molluscs, and are transported to these small animal forms. The small forms are eaten by successively larger ones until the radioactive materials are assimilated by fish. The fish may be considered not only the ultimate link of the aquatic food chain but also a connecting link to the land vertebrates...P^{32} is thus concentrated in aquatic organisms on the order of 10^5 [100,000] times...Organisms...thus built up to a concentration in the tissues which might be thousands of times that present in the water. This obviously would make the organism appreciably radioactive.[24]

In 1950, Parker reported that river fish "sampled at Hanford during this quarter [April–June 1950] were found to contain on the order of three times more activity than was found during the same quarter in 1949."[25] He found similar increases the next quarter.[26] In a 1949–1950 study, pekin ducks showed that "activity densities were generally significantly higher than those observed" in 1948.[27] Researchers noted a "general increase in the activity density [in river water] from gross beta emitters at nearly all river sampling locations."[28]

In December 1950, according to Parker, "the activity density of plankton reached an all-time high...in the vicinity of Hanford...observed values were two to four times higher than during 1949."[29] Two HI scientists reported that "the actual specific concentration factor of P^{32} may be well over one million for some organisms at Hanford." Ominously, they stated that "abnormalities...though they may initially affect only the most minute and seemingly insignificant organisms, might produce indirect effects of considerable magnitude which could reflect upon...the welfare of man."[30]

In 1951, HW data showed radioactivity levels in river water at Hanford, Richland, and Pasco-Kennewick had increased by an average of ten-fold from 1947 through mid-1951.[31] Parker reported "nearly a 3-fold increase over values obtained in 1950, and a 5-fold increase over 1949."[32] He blamed "changes in pile [reactor] operation."[33] In 1952, he reported that "activity density in all aquatic organisms were from 30–75% higher than those observed one year ago."[34] Measurements of beta activity emitters in reactor effluent entering the Columbia River at the outlet of the retention basins showed increases from nine-fold to 18-fold over 1947.

In 1953, Parker reported that "whitefish [near Hanford]...contained significant amounts of radioactivity."[35] By autumn, radioactivity in plankton had increased 25 percent over early 1952, and algae had tripled in activity levels. Caddis fly larvae had increased in activity by two and one-half times since autumn 1952.[36]

In August 1954, about 8,000 curies per day of radioactive material were being poured in the river by Hanford reactors. "The most significant hazard," Parker stated, "arises from sport fishing...Presently, the consumption of about 35 pounds of whitefish per week is the appropriate limit for man, in the absence of all other radiation hazards. Individual fish reach contamination levels appreciably higher than the average on which this limit is based... There are many secondary hazards to man, such as waterfowl contamination or uptake in irrigated crops."[37] However, the AEC's Division of Biology and Medicine disagreed with Parker's calculations, finding that the actual limit of radionuclide ingestion of P^{32} in man "would limit the average consumption of such fish [whitefish] to about 7 lbs/wk [pounds per week]. This is less by a factor of 5 than the value suggested by Dr. Parker...it appears quite reasonable to assume that the quantity of fish eaten by persons who like to fish and to eat fish may average several pounds per week."[38]

In January and June 1955, respectively, the KW and KE Reactors started operating. Known as the "jumbos," the addition of these reactors essentially doubled plutonium production at Hanford. Together, their effluents added approximately 140,000 extra gallons per minute of contaminated liquids to the Columbia River.[39] In addition, by the end of 1955, the holdup time of reactor effluent water in the retention basins had shrunk to between 1.1 and 3.6 hours.[40] That year, combined reactor effluent discharges averaged 456,000 gpm. An HW report showed that effluent was 0.24% of total river flow in January 1948, but was 1.51% of total river flow in December 1955, an increase of six-fold. "There has...been a pronounced over-all increase in the radioactive contamination levels in the organisms during the years [1948–1955]," the report continued, "which is quite comparable to the increase in total reactor power levels."[41]

That year, Hanford scientist H. Kornberg reported that "compared to the concentration of P^{32} in the Columbia River water, small fish contain forty thousand to three hundred thousand times more P^{32} per unit mass."[42] Colleague M. Mickelson noted that "the P^{32} has increased at a rate faster than that expected from a proportionality with [reactor] power level...there is [today] a sufficient concentration in some fish at certain seasons to furnish a near permissible radiation dose to anyone eating them in quantity."[43] In 1956, scientist P. McMurray reported that "assuming 100% absorption of P^{32} by the human and current effluent conditions, the amount of whitefish consumed per person per week which will currently give the permissible body burden is estimated to be 8 pounds."[44] That autumn, two HI scientists reported that "the rate of consumption necessary to produce maximum permissible limits from fish caught outside the [Hanford] reservation during the fall of 1956 was 2.7 pounds per week."[45] At that time, three HI scientists estimated that "about 10^4 [10,000] curies per day of beta emitters...are discharged from the eight reactors to the river."[46] Another reported that average amounts of curies discharged to the Columbia in reactor effluent rose from 7,500 curies per day in 1954 to 13,000 curies per day in 1957.[47]

On February 18, 1958, the Hanford reactors set a record for eight-reactor operations by running continuously, with no shut-downs or interruptions, for 8.9 days.[48] Later that year, HW scientist B. Andersen again acknowledged that "the increase in radioactive materials to the river was the result of higher power levels."[49] By the end of 1958, "the average contamination in whitefish from Priest Rapids was from four to ten times the values for last year."[50] In January 1959, power levels again were increased at all eight Hanford production reactors, making a combined total of 19,600 MW.[51] In late 1959, Hall reported that "the criterion for phosphorus-32 is being exceeded at this time."[52]

In early 1960, an HW report stated that approximately 13,300 curies per day of beta-emitting radionuclides, measured after four hours of decay, were being released to the Columbia River by Hanford reactors. Some exceptions occurred when average amounts of release fell slightly below 10,000 curies per day, and on a few occasions, averages reached nearly 20,000 curies per day. During April 1959, the daily average peaked at 20,300 curies per day.[53] During the period June 1957 to November 1959, the number of curies of P^{32} discharged into the river by Hanford reactors rose from an average of 21.3 per day to 90.4 per day, an increase of more than four-fold. The number of curies of As^{76} discharged during the period

June 1957 to December 1958 rose from a daily average of 323 curies to 999 curies, an increase of more than three-fold. The number of curies of Zn^{65} discharged during the period January 1958 to November 1959 rose from a daily average of 53.3 curies to 152.2 curies, an increase of three-fold.[54] Another HW report at the same time stated that the maximum permissible concentration (MPC) of P^{32}, per national and AEC standards, "could be obtained from the consumption of about one pound of whitefish flesh each week...the P^{32} content of some ducks which are killed along the river near Hanford is similar to that found in the fish."[55]

RIVER AND FISH STUDIES (RADIOLOGICAL) AND RESULTS, 1961–75

In 1961, power levels again were increased at all eight operating Hanford production reactors. The levels totaled 21,560 MW, as opposed to the 1959 total of 19,600 MW.[56] That summer, an HW report estimated that in the previous year the amounts of curies per day in river water traveling past Pasco (a town about 30 miles south of the furthest downstream reactor) increased from 43 to 69 for P^{32} (an increase of 62 percent), from 400 to 430 for As^{76}, and from 700 to 800 for Np^{237} (both latter increases of about nine percent).[57]

Throughout 1962, the HW reactors continued to operate at peak power and temperature levels. In October, coolant water flow increased by 30,000 gpm at each of the K Reactors.[58] During 1963, the Hanford reactors set several production records, both for sustained input and continuous operation.[59] That year, the city of Richland (near Pasco) began using Columbia River water for sanitary and drinking water, instead of using water from wells that had been in service since World War II. Foster stated early in 1964 that "the outlook is for an annual exposure to the GI tract of Richland residents approximating four times that now experienced at Pasco...This higher dose for Richland is due primarily to less radioactive decay in the river and also to shorter hold up time in the water plant."[60]

On December 31, 1963, Hanford's N Reactor began operating. It was the largest "weapons production reactor in the U.S. arsenal, capable of producing more plutonium, more economically."[61] For the next 12 months, nine production reactors operated at Hanford. On January 7, 1964, President Lyndon Johnson announced that the Hanford production reactors would close in phased sequence beginning late that year.[62] Nevertheless, the Hanford reactors continued to set production records throughout most of 1964. In December, the 1964 "input production record exceeded record year 1963 by 8.2%."[63] Concentrations of P^{32} in whitefish increased at the same time.[64]

In December 1964, DR Reactor shut down, and H and F Reactors were deactivated in April and June 1965, respectively. Nevertheless, in March, and again in May 1965, the total monthly shipment (output) of irradiated metal from all Hanford production reactors set new high records.[65] Contamination in the effluent water of the operating Hanford reactors climbed throughout the year, showing average increases of P^{32} content at B, D, and C Reactors of 19 percent over 1964, and average increases of As^{76} content of nearly 50 percent over 1964. At the K Reactors, P^{32} concentration in the effluent rose by two and one-half times over that of 1963, and As^{76} concentration increased by seven times over that of 1963.[66] In September 1965, an operating accident at the Plutonium Recycle Test Reactor (not a production reactor) in Hanford's 300 Area, just north of Richland, added contamination to river water and "the corresponding radiation doses from the consumption of sanitary water."[67]

During 1966, five production reactors operated at HW, and continued to set high production records for themselves, although total production records never reached those of 1964.[68] From July 9 to August 25, 1966, all of the Hanford production reactors were shut down due to a strike by the operating union at HW. This shutdown period, HW scientists reported, "produced marked results in the concentrations of radionuclides in the biota. Concentrations of radionuclides decreased rapidly to varying levels. Chromium-51 and phosphorus-32 decreased two or three orders of magnitude... Decreases of an order of magnitude...were common in most organisms." After the shutdown period ended, "all organisms rapidly accumulated the radionuclides following resumption of reactor operation and near equilibrium concentrations in most organisms were approached within two or three weeks."[69]

In June 1967, D Reactor shut down, bringing the volume of effluent water being discharged from Hanford's production reactors to approximately 65 percent of the 1962–64 peak. Nevertheless, B, C, and

K Reactors again set records for efficiency, production, and continuous operation.[70] Average concentrations of P^{32} in the flesh of whitefish sampled downstream from the Hanford reactors was higher in 1967 by a factor of nearly two and one-half times that of fish caught in 1966. HW scientists attributed the lower concentration in 1966 "to the extended reactor shutdown during July and August."[71]

In February 1968, B Reactor was deactivated, bringing the volume of reactor effluent entering the Columbia River to about 55 percent of the 1964 high levels. The C and K Reactors continued to set individual production records that year. In April 1969, C Reactor shut down. KW Reactor set efficiency records that year, while KE Reactor operated normally.[72] In January and April 1971, the KW and KE Reactors, respectively, shut down. N Reactor closed briefly but was restarted later in the year. During the 1970s, some of N Reactor's secondary cooling systems were converted from once-through to recirculating cooling systems, further reducing effluent discharge to the Columbia River.[73] By the late 1970s, the volume of effluent discharges to the Columbia River stood at less than two-tenths of one percent of the 1964 high levels.

In 1976 and 1980, Hanford studies, now public, showed that, for the most part, radionuclide burdens from Hanford production reactor effluent in the Columbia River water returned to "essentially undetectable levels within 18 to 24 months of cessation of input of once-through cooling water in the river…The biota in the Columbia River ecosystem below Hanford are being exposed to…much lower, ambient level[s] of radionuclides in their environment."[74] A 1976 study stated that "long-lived radionuclides resulting from past Hanford operations are currently measurable along the Columbia River shoreline, islands, bottom, and slough areas. These radionuclides are associated with river sediment."[75]

Effects of Fuel Ruptures on Hanford River Water and Organisms

Effluent discharges from normal Hanford production operations weren't the only problem to beset the Columbia River. Failures of the metal coatings covering uranium fuel elements (known as "slugs") added extra radionuclide burden to the discharge water. "Hot spot" ruptures (single point failures that produced small holes in the fuel coating) were less serious than "side-split" (or simply "split) ruptures wherein the entire length of a fuel element's coating opened up. Split ruptures deposited more radionuclides into effluent water and posed a greater chance of the fuel element becoming stuck in the reactor. No fuel element failures occurred at Hanford during World War II. However, by December 1945, 125 slugs with "blisters" had been found, and 400 more blistered slugs were noted by spring 1946.[76] Two outright fuel failures occurred in 1948, and GE Hanford Co. built "earth reservoirs" (trenches—known at Hanford as "cribs") to "stop the flow [of reactor effluent from fuel failures] to the river and divert it." The cribs would allow effluent to percolate slowly through the ground to the river, providing some time for radioactive decay.[77]

Beginning in 1950, fuel failures became a significant source of contamination entering the Columbia River from Hanford reactors. In 1951 there were 115 fuel failures and 142 in 1952.[78] At that point, Parker wrote that "we are much concerned about the quantities of fission products, as differentiated from activation products which go into the river…fission products may be significantly more hazardous per rep than are the particular activation products in reactor effluent."[79] The year 1955 brought fuel failures at Hanford to the all-time high number of 242.[80] However, by 1959, changes in fuel element manufacturing and coatings had brought the rupture problem down somewhat. That year saw only 71 ruptures, but a Hanford study stated that "rupture debris may contribute appreciably to the overall contamination of the river… Estimates for the following years [after 1954] showed the quantity of rupture debris increasing materially as a result of greater frequency of the more severe failures… The present rupture frequency approximates one rupture each two days… A very unlikely coincidence of [multiple] ruptures could raise the river to three times the occupational limit (MPC—maximum permissible concentration, as specified by national and AEC standards)."[81]

In 1960, the number of fuel failures rose to 130, but declined nearly steadily to 89 by 1965.[82] Although the numbers went down, fuel ruptures continued to pose important contamination risks, including causing two of the 14 most serious operating accidents in Hanford's history. At the KW Reactor, fuel overheated and ruptured in a tube without water flow in 1955. The tube also ruptured, damaging nearby reactor blocks and other components and resulting in a 10-week reactor shutdown, with repairs costing $550,000 in 1955 dollars (nearly $5 million today). It was the most expensive repair in the history of Hanford production reactors.[83] A second serious fuel failure occurred at KW Reactor on June 19, 1968, necessitating six weeks of shutdown and repair.[84] Other salient consequences occurred when ruptured fuel elements burned upon discharge, contaminating not just the Columbia River but the atmosphere. Such incidents occurred at H Reactor in 1955 and KW Reactor in 1959.[85] When multiple fuel elements ruptured at once, consequences also became more severe, as happened at H Reactor in 1954 and 1957, DR Reactor in 1957 and 1958, and C Reactor twice in 1962.[86] Other serious situations occurred when fuel failures were not detected quickly. One such incident at N Reactor in 1973 resulted in concentrations of some radionuclides in the Columbia River of 100 times the normal concentrations, and in discharges to the atmosphere.[87] Ruptures of special test or target elements, or of enriched elements, also had grave consequences, as when an experimental fuel element ruptured in the KE Reactor on May 12, 1963, releasing about one pound of uranium into the Columbia River. It was the largest single release of fission products to the river in Hanford's history.[88] A similar, albeit less serious, incident occurred at KW Reactor in 1966.[89] Other important consequences occurred when process tube ruptures accompanied fuel ruptures, as occurred multiple times.[90] In total, more than 2,100 fuel failures occurred in Hanford production reactor history.

EFFECTS OF CHEMICALS IN HANFORD REACTOR EFFLUENT DISCHARGES

Aside from radioactive discharges, Hanford's reactors delivered major chemical burdens to the Columbia River. The chemicals were added to the influent water of the reactors to preserve the metal process tubes, inhibit algal growth, regulate pH, and perform other functions. The major chemicals were:

- Ferric or aluminum sulfate and activated silica as coagulants
- Chlorine to inhibit growth of algae
- Lime to regulate pH
- Sodium dichromate (containing hexavalent chromium, expressed as Cr^{+6} or CrVI) to inhibit corrosion[91]

The pH of Columbia River raw water varied between 7.6 and 8.4, a level that early Hanford scientists feared would cause corrosion of the aluminum process tubes and the aluminum-silicon cladding that covered the uranium fuel elements. Therefore, they added 1.8–2.2 parts per million (ppm) sodium dichromate ($Na_2Cr_2O_7$), which flowed into the tiny cracks and crevices of the aluminum surfaces and acted to passivate the metal. The interaction between pH and the hexavalent chromium (Cr^{+6}) was crucial, as the scientists believed that a pH much below 7.0 would reduce the Cr^{+6} to trivalent chromium (Cr^{+3}) and lay down a heavy film on the process tubes.[92]

Trouble was noted by 1948, when HW scientist P. Olson stated that reactor cooling water contained "toxic chemicals...which might poison the fish or their food organisms."[93] The following year, Foster also reported that some of the water treatment chemicals "are toxic to aquatic organisms when present in significant amounts. The chronic poisoning of organisms by such substances produces symptoms which are not unlike effects which might be expected from radiation damage."[94] In April 1952, Hanford scientists tested eliminating sodium dichromate to reactor process water but reinstituted it within a year due to corrosion effects in the reactors.[95]

In 1954, Parker acknowledged that "there are problems, or potential problems arising from chemical toxicity...in the river."[96] By this time, the dichromate was recognized as the primary chemical of concern in reactor water, not only because it was harmful to fish directly, but because it caused the radioisotope Cr^{51} to form in the effluent.[97] That same year, Foster and two colleagues stated that "chemical toxicity rather than radioactivity" was among "the most deleterious to the local salmon populations... sodium dichromate is the most toxic component of the process water."[98] In 1955, HW scientist Mickelson stated that "temperature and chromate ion concentrations of reactor effluent were observed to be more damaging to salmon eggs than its radioactive isotope content; existing chromate ion concentrations in the river closely approached permissible levels for the maintenance of the fish population.[99] In spring 1957, Foster recommended that the "monthly average concentration of Cr(VI) in the Columbia River [should] not exceed 0.02 ppm...The fish can apparently tolerate this concentration indefinitely without significant inhibition, but this limit does *not* contain a factor of safety...chromium is a toxic substance and the quantities discharged into the river warrant close attention." In some cases, concentrations of hexavalent chromium as high as 0.016 ppm already had occurred.[100] In 1961, General Electric Hanford Co. reported that measurements in river water at F Reactor's intake showed increases in CrVI content from 0.006 ppm in November 1954 to 0.020 in January 1961, an increase of more than three-fold.[101] At Pasco, CrVI content of river water increased from 0.006 ppm in April 1960 to 0.016 in March 1961, an increase of nearly three-fold in 11 months.[102]

A distinct part of the chemical burden in the Columbia River from Hanford production reactors came from reactor purges, in which films that built up on reactor piping were periodically removed. Inside the reactor, films could impede water flow and interfere with neutron bombardment in the basic fission reaction. On the downcomer, outlet piping, and retention basins, films caused increased radiation dose to workers because they collected nuclides in concentrated form. By 1945, HW operators were purging the reactors with a diatomaceous earth slurry which became the standard at Hanford (with some changes in trademarked slurry products over the years).[103] By 1952, as reactor power levels and temperature of reactor effluent increased, films built up more readily on reactor piping, and each reactor needed to purge about once a month. HW operators began using operating purges, known as "hot" purges because they occurred while the reactor was running. Hot purges were very effective in removing reactor films, but "produced a five-fold increase in effluent activity over 'cold' purges (those conducted while the reactor was shut down)."[104] During 1953 through 1955, several different purge agents were tried in operating purges. Studies in 1956 showed that activity levels in purge effluent were higher than activity in "regular" effluent by an average factor of 4.5, but specifically higher by a factor of 11.5 for P^{32} and a factor of 28.3 for Zn^{65}.[105] A study in 1957 showed that operating purge effluent could contain as much as 25 times the P^{32} and 20 times the strontium isotopes as regular effluent.[106] That year, with reactors purging more than once a month and sometimes as frequently as twice a month, HW restricted operating purges to only days when the river temperature was below 15°C (59°F). Additionally, only 10 purges were allowed during the period of low river flow from August 15 to November 1.[107]

EFFECTS OF RIVER TEMPERATURE RISES AS A RESULT OF
HANFORD REACTOR OPERATIONS

Still another affront to the Columbia River came from the heat load imposed by the heated water discharged from the reactors. Bulk outlet water temperatures limits for the Hanford reactors began at 65° Centigrade (C) (149° Fahrenheit [F]) in WWII; stood at 80°C (176°F) by 1951; were raised to 90°C (194°F) in 1952; stood at 93°C (199°F) in 1960; and at 95°C (203°F) in 1964 (and onward).[108] As early as 1946, Hanford scientists noted that radioactivity accumulated faster in fish tissues when the water was warmer.[109] In 1951, Parker noted that experiments demonstrated that "high water temperatures favored disease organisms... the importance of temperature in contributing to [fish] mortality is indicated."[110] In 1952, he wrote that "current disposal practice increases the river temperature to a point which is conceivably critical in the month of October [a time of low river flows]...future increments of power level will aggravate the condition."[111] Two years later, he told the AEC that "there are problems, or potential problems arising from...thermal effects in the river."[112] That same year, Foster co-authored a report stating that "the effluent discharged into the river is on the order of 50°C [122°F] higher than the river temperature...increased river temperatures constitute the greatest potential threat to the local salmon...[We found] a marked increase in mortality in fish reared at temperatures kept approximately 5°C above normal for the Columbia River."[113]

In 1956, another Hanford report revealed that "the current increase in river temperature due to HAPO effluents is about 1°C...an increase of 4°C in the local spawning area may result in the mortality of about 50% of the salmon eggs."[114] That same year, a study showed that a 2°C rise in river temperature above normal resulted in a 5 percent increase in the mortality of whitefish eggs. An increase of 3°C resulted in an 8 percent increase in mortality of the eggs.[115] A 1957 study showed that a 2°C rise in river temperature above normal resulted in an 8 percent increase in the mortality of whitefish eggs. An increase of 3°C resulted in a 21 percent increase in mortality of the eggs.[116] In 1958, Foster wrote that "valuable species of Columbia River fish, and especially the fall run of chinook salmon, are definitely vulnerable to further temperature increases in river."[117] The following year, Hall stated that the "opinion of the Biology

Operation [of HW is] that...the probability of an epidemic of disease among salmon increases with temperature over 20°C and with the length of time that such temperatures are maintained...in 1958 epidemic disease was reported when temperature between 19°C and 21.5°C persisted for nearly 10 weeks."[118]

That summer, HW began a program to "artificially" cool the Columbia River water during the months of low river flow and warm weather. The AEC arranged to have very cold water from the bottom of Lake Roosevelt (the reservoir on the upstream side of the Grand Coulee Dam) released into the Columbia River to flow through and by HW. The program ran from August 6, 1959, until October 5, 1959. The temperature reduction in the Columbia at Hanford was estimated to be 2°C [approximately 4°F].[119] The cooling program continued each year through the summer of 1965, after which time reactor shutdowns made it unnecessary. It cooled the Columbia's water in the vicinity of Hanford reactors by an average of 1°C, and a maximum of 2.7°C. The lowered inlet water temperatures allowed higher power levels in reactor operations, while still not exceeding maximum desired bulk outlet temperatures.[120] In 1962, in another attempt to reduce heat load in the Columbia River, Hanford scientists investigated cooling towers for the reactors. They concluded that "direct-cycle, induced draft cooling towers would be effective for reducing the gross heat load of the reactor effluent, but...The major problem...would be radioactive contamination control of the immediate vicinity. In addition to the water evaporated...mist and spray would certainly contaminate the area surrounding the towers."[121] The concept was studied again in 1967, but again dismissed due to "contaminated fog and spray" issues.[122]

REACTOR EFFLUENT DISPOSAL ALTERNATIVES

In the mid-1950s HW scientists began to discuss alternative effluent disposal methods. Production was rising, the mission seemed vital, the choice to stop was nonexistent, and yet they knew the magnificent river was suffering. It had a limit—as all of nature does. They began an unabridged search for a different way. In 1955, scientist J. Honstead proposed creating an "inland lake system," wherein effluent would be diverted through a canal from each reactor to an artificial lake system on either side of Gable

Mountain (about three miles inland). The effluent would percolate slowly through the soil and groundwater back to the river, while the routing and travel time would reduce the radioactivity and heat levels to safer levels. Then the effluent could be discharged through another canal on the river bank.[123] Hanford scientist M. Mickelson concurred with the plan: "Such a system could mitigate river pollution problems by permitting thermal cooling of the effluent, allowing decay of radioisotopes of significance to GI [gastrointestinal] tract irradiation from downstream drinking water usage."[124] However, a Hanford meteorologist quickly noted that steam fog capable of producing stratus clouds would be formed, and radiation fog could be generated, even on cloudless days, "over an extensive area," affecting plants and vehicular traffic. Radioactive frost, dew, and rime would be deposited in the vicinity of the lakes and canals, pushing the loadings on electrical distribution systems, and heating and ventilation facilities "beyond design" capacities.[125]

In 1962, the inland lake or "artificial lake" concept was revived.[126] Such a "system [would] provide some degree of decontamination...[and] a significant holdup time for decay of the short-lived (less than two day half-life) radioactive isotopes but will have little effect on the long-lived isotopes."[127] However, drawbacks again included radioactive fog and other issues: "The increased height of the water table that may result might affect structures in some of the reactor areas...[and] might cause springs to develop in the lower areas...There is [also] a possibility, however, that saturation of [soils and clays]...could be approached, in time reducing the adsorption capacity...The greatest problem area appears to be associated with the effect of the canal and lake system on the underground water table. Undoubtedly it will rise."[128]

Despite these drawbacks, a 600-foot trench to test a ground disposal system was placed into service at the F Reactor Area in 1964. However, a pronounced mound soon developed in the shallow groundwater table beneath the trench, and springs developed along the perimeter road. After F Reactor shut down in 1965, HW scientists developed a concept for a canal diversion system that would route effluent from the B and C Reactors through a trench to the KE-KW Reactor Area, and then send the combined effluent of these four reactors through a larger ditch to a river discharge point just downstream of the F Reactor Area. The total length of the canal would allow many hours of radionuclide decay time but would contaminate groundwater all along its route and form "polluted...swamp areas" at the discharge point. This idea was not pursued beyond the design stage, but a trench with a different orientation was placed into service in the B Reactor Area on October 30, 1967. However, by early 1968, an "increase in the level of the water table in the vicinity of B Area is apparent...[and] extensive new seepage areas...formed along the riverbank." The shutdown of B Reactor that February ended the trenching test.[129] In 1967, HW scientists estimated the costs of building various types of inland lakes, canals, and trenching systems to dispose of production reactor effluent at between $6.5 million and $20.5 million ($46 million to $145 million in 2019 dollars).[130] They also investigated modifying and using the Hanford Irrigation Ditch from the pre-site town of Hanford to dispose of reactor effluent, and placed the capital cost for modifications at $12 million (about $85 million in today's dollars).[131]

In 1959, HW scientist J. Healy explored the radical idea of simply eliminating use of the reactor retention basins for ordinary effluent. In this concept, the basins would be used only to retain debris from fuel ruptures and reactor purges. Basin retention time was already so brief, he reasoned, that the basins did little to reduce the entry of "'longer lived radionuclides' (including P^{32}) into the river." Eliminating the use of retention basins, he said, would increase "radiation doses to organisms in the vicinity of the outlet...but...there would be little or no effect upon the quantities of long-lived isotopes available to the food chains in the river...[This plan] leaves the possibility of the production of a small area around the outlet which will be denuded of organisms."[132]

HW scientists also explored varying effluent treatments. In 1958, they tried passing reactor effluent through beds of various metals, metal oxides, and ion exchange resins to entrap certain radionuclides. A small-scale study that year found that steel wool and several anion exchange resins worked best, and that aluminum, steel, and iron minerals removed up to 90 percent of the P^{32} and As^{76}.[133] In 1959, in a larger test, a laboratory-sized bed of aluminum cuttings and turnings reduced P^{32} by 43 percent and As^{76} by 65 percent. Corrosion levels were high in the aluminum.[134] In 1960, a pilot-scale test bed, 20 feet long, was installed in a steel-plated tank near a D Reactor retention basin. The bed was filled with shavings of aluminum, and part of D Reactor's effluent was diverted through it. After eight months of operations, so much algae had collected in the front end of the bed that the flow rate had to be decreased to the point that made the system impractical for the HW reactors, each of which produced 80,000–115,000 gallons of effluent per minute. Additionally, so much radiation built up on the tank that HW scientists realized that a full-scale facility would need complete shielding. Importantly, corrosion was measured at a rate that would necessitate "virtually total replacement" of the apparatus after just two and one-half years. As a result, the idea of decontamination of reactor effluent via aluminum beds was abandoned in 1961.[135]

Next, scientists tried varying the influent water treatments. Something had to work to reduce the amount of reactor radionuclides discharged to the great river. They wanted to reduce or remove parent elements from the process water so that radioisotopes could not form from them. In a 1960 test, a 90-percent ionic reduction in the elemental arsenic and phosphorus content of filtered water was achieved when aluminum nitrate at 25 ppm was used as the flocculent in the filtration process. Because this treatment was very expensive, it was considered for possible use only during the August to October "critical season" for P^{32}, the time when low river flows helped dramatically raise the average P^{32} levels in whitefish.[136] In 1961, a full "high alum feed" experiment was started at F Reactor, and then implemented at all reactors by mid-July. Aluminum sulfate quantities were increased in influent water from double to more than quadruple amounts. Results showed a 40 percent reduction in the P^{32} content of reactor effluent, but "ledge corrosion" and film build-up in process tubes showed a "marked increase."[137] The primary radionuclides in the films that accumulated during the high alum feed program were P^{32}, Fe^{59}, Zn^{65}, Cr^{51}, Co^{60}, and Sc^{46}.[138] Nevertheless, the program was adopted at all of the HW reactors in 1962, but with modifications lowering the aluminum sulfate levels by varying degrees among the reactors in the ensuing years.[139]

Many new protective coatings and process water additives also were tested. The organic coatings that were tried all decomposed when subjected to neutron flux within the reactors. Anodic coatings and electropolishing of the tubes actually increased film build-up, and some silicon resins and ordinary commercial inks were effective as tube coatings, but cost and various in-reactor difficulties prevented their large-scale use. Hanford scientists also learned that adding sodium silicate to the process water reduced P^{32} and As^{76} concentrations in effluent, but the sodium silicate added an undesirable chemical load to the Columbia River.[140] By the mid-1960s, as the reactors aged, the corrosion situation became even worse. A 1964 experiment showed that deionization of raw river water to remove minerals and mineral salts substantially reduced concentrations of P^{32} and other radionuclides of concern, especially when used in combination with process tubes made primarily of zirconium.[141] However, retubing the aging HW reactors was a risky operation, because it usually involved some degree of graphite loss or damage, and thus further weakened the mechanical strength of the graphite cores and shortened reactor life. Additionally, installation of large-scale deionization plants at the HW reactors was not deemed economically feasible, as reactors were shutting down.

Hanford scientists knew that the most effective means of reducing the radionuclide burden to the Columbia River from HW reactors would be to convert the once-through cooling systems to recirculating systems.[142] In 1958, they studied this possibility for the K Reactors.[143] In 1967, they again evaluated conversion of the K Reactors to recirculating water cooling systems, estimating the capital cost at approximately $32 million for each reactor.[144] Considering the high cost alongside changing defense needs and political priorities, the conversions were never accomplished. That same year, essentially out of alternatives, HW scientists even investigated installing a log boom in the Columbia River "around the area of current effluent discharge, encompassing the effluent plumes, with the intent of controlling access to the plumes until dilution has taken place." The log boom was planned to be 10–15 miles in length but was never built.[145]

In summary, none of Hanford's efforts to reduce the radionuclide, chemical and heat loads to the Columbia River, while still operating reactors whose technology was primitive by the standards of the late 1960s and early 1970s, was effective. Time had passed by these hulking concrete behemoths. In the end, as it has done throughout millennia, the Columbia healed itself, reengineering through time what the engineers and scientists had tried to subdue.

Notes

1. M. S. Gerber, "History of 100 B/C Reactor Operations, Hanford Site," WHC-EN-RPT-004, Westinghouse Hanford Co. (Richland, WA), April 1993.

2. Research and Engineering Section, N Reactor Department, "100-N Technical Manual," HW-69000, General Electric (GE) Hanford Co. (Richland, WA), March 5, 1963; B.C. Hopkins, "N Reactor Plant Manual," UNI-M-94, United Nuclear Industries, Inc. (UNI) (Richland, WA), March 29, 1979.

3. D. L. DeNeal, "Historical Events Reactors and Fuels Fabrication," RL-REA-2247, GE Hanford Co., July 1, 1965; D. L. DeNeal, "Historical Events—Single Pass Reactors and Fuels Fabrication," DUN-6888, Douglas United Nuclear Co. (DUN) (Richland, WA). April 10, 1970; J. W. Talbot, "Operating History of Hanford Piles," HW-37304, GE Hanford Co., June 15, 1955; R. E. Trumble, "Hazards Summary Report—Projects CG-558 and CG-600 Reactor Plant Modifications, Volume I," GE Hanford Co., April 2, 1957.

4. DeNeal, RL-REA-2247; DeNeal, DUN-6888.

5. DeNeal, DUN-6888, 49; J. W. Ballowe, "Summary of Alternate Methods of Reactor Effluent Treatment and Disposal," DUN-2231, DUN, March 22, 1967.

6. R. B. Hall and C. Jerman, "Effect of Proposed HAPO Reactor Production Increases on Radioactivity Discharged to the Columbia River," HW-64517, GE Hanford Co., March 29, 1960, 2.

7. DeNeal, DUN-6888, 49.

8. R. B. Hall and C. Jerman, "Reactor Effluent Water Disposal," HW-63653, GE Hanford Co., February 1, 1960, 2–3.

9. J. J. Davis, D. G. Watson, and C. C. Palmiter, "Radiobiological Studies of the Columbia River Through December 1955," HW-36074, GE Hanford Co., November 7, 1956, 14.

10. C. C. Gamertsfelder, "Effects on Surrounding Areas Caused by the Operations of the Hanford Engineer Works," HW-7-5934, GE Hanford Co., May 11, 1947.

11. Note: all reports cited in this paper were originally classified Secret, unless otherwise noted.

12. Hall and Jerman, HW-63653, 2; Foster, R. F. and R. L. Junkins, "Off-Project Exposure from Hanford Effluent," HW-63654, GE Hanford Co., February 1, 1960, 3.

13. Hall and Jerman, HW-63653; Clukey, HW-54243; R. F. Foster, "Radiobiological Problems in the Columbia River, Draft" HW-23793. GE Hanford Co., April 1949; D. G. Watson, and J. J. Davis, "Concentration of Radioisotopes in Columbia River Whitefish in the Vicinity of Hanford Atomic Products Operation," HW-48523, GE Hanford Co., February 18, 1957.

14. F. T. Matthias, "Journal and Notes, 1943–46," Diary held in U.S. Department of Energy Archives, February 17, 1945, March 6–7, 1945, March 13, 1945, May 12, 1945; R. F. Foster, interview in *Oregonian* (Portland, OR), May 12, 1986, 1; A. Olson, "Fish and Fish Problems of the Hanford Reservation," HW-11642, GE Hanford Co., November

23, 1948, 2; L. R. Donaldson, "Problems of Liquid Waste Disposal," UWFL-15 (CIC-50004), University of Washington Applied Fisheries Laboratory, November 10, 1948, 1–2; R. F. Foster, J. J. Davis, and A. Olson, "Studies on the Effect of the Hanford Reactors on Aquatic Life in the Columbia River," HW-33366, GE Hanford Co., October 11, 1954, 3.

15. M. S. Gerber, *On the Home Front: The Cold War Legacy of the Hanford Nuclear Site*, (Lincoln, NE: University of Nebraska Press), 1992, 1997, 2002, and 2007, 117.

16. J. W. Healy, "Accumulation of Radioactive Elements in Fish... in Pile Effluent Water," HW-3-3442, Hanford Engineer Works (Richland, WA), February 27, 1946; K. E. Herde, "Studies in Accumulation of Radioactive Elements in Chinook Salmon Exposed to a Medium of Pile Effluent Water," HW-3-5064, Hanford Engineer Works. October 14, 1946.

17. K. E. Herde, "Radioactivity in Various Species of Fish from the Columbia and Yakima Rivers," HW-3-5501, GE Hanford Co., May 14, 1947, 1, 2, 11.

18. K. E. Herde, "A One Year Study of Radioactivity in Columbia River Fish," HW-11344, GE Hanford Co., October 25, 1948, 2, 7. The concept of "tolerance dose" is a historical term used in the new science of health physics in the 1930s and 1940s. It referred to the attempt to find radiation doses from various radioactive materials that would be permissible to a receiving bioorganism without producing harmful effects. Herbert Parker brought this term to Hanford in the site's early years. The term and concept are no longer used. Millirep means one-thousandth of a rep. Rep is a historical term that is an acronym for "roentgen equivalent physical." It was an early attempt to define a radiation dosage received by an organism. It was replaced by the dosage term rem (roentgen equivalent man).

19. Ibid., 2, 4.

20. R. W. Coopey, "Preliminary Report on the Accumulation of Radioactivity as Shown by a Limnological Study of the Columbia River in the Vicinity of the Hanford Works," HW-11662, GE Hanford Co., November 12, 1948, 1–2, 11.

21. Donaldson, UWFL-15 (CIC-5004), 4–5.

22. Herde and Cline, HW-12079, 3, 5.

23. W. Singlevitch and H. J. Paas, "Radioactive Contamination in the Environs of the Hanford Works for the Period October, November, December 1948," HW-13743, GE Hanford Co., June 22, 1949, 3, 20–21

24. Foster, HW-23793, 6, 8, 9, 3.

25. H. M. Parker, "Quarterly Progress Report, Research and Development Activities for April–June 1950," HW-18371, GE Hanford Co. July 19, 1950, 9.

26. H. M. Parker, "Quarterly Progress Report, Research and Development Activities for July–September 1950," HW-19146, GE Hanford Co. October 16, 1950, 8.

27. K. E. Herde, R. L. Browning, and W. C. Hanson, "Activity Densities in Waterfowl of the Hanford Reservation and Environs," HW-18645, GE Hanford Co., August 21, 1951, 6–7, 9.

28. H. J. Paas, "Radioactive Contamination in the Environs of the Hanford Works for the Period October, November, December 1950," HW-21566, GE Hanford Co., July 13, 1951, 4.

29. H. M. Parker, "Annual Report of the Health Instruments Division, 1950," HW-21699, GE Hanford Co., July 20, 1951, 35.

30. J. J. Davis and C. L. Cooper, "Effect of Hanford Pile Effluent Upon Aquatic Invertebrates in the Columbia River," HW-20055, GE Hanford Co., January 19, 1951, 32–34, 42–43, 61.

31. "Concentration of Gross Radioactivity in Columbia River Water," GEH-18774 (also identified as AEC-38104 and HAN-38104), U.S. Atomic Energy Commission (AEC) (Richland, WA), 1951.

32. H. M. Parker, "Quarterly Progress Report, Research and Development Activities for January–March 1951," HW-20866, GE Hanford Co., April 18, 1951, 9.

33. H. M. Parker, "Quarterly Progress Report, Research and Development Activities for July–September 1951," HW-22576, GE Hanford Co., October 1951, 4, 7–8.

34. H. M. Parker, "Quarterly Progress Report, Research and Development Activities for January–March 1952," HW-24131, GE Hanford Co., April 16, 1952, 7.

35. H. M. Parker, "Quarterly Progress Report, Research and Development Activities for October–December 1952," HW-26523, GE Hanford Co., January 5, 1953, 5–6.

36. H. M. Parker, "Quarterly Progress Report, Research and Development Activities for July–September 1953," HW-29519, GE Hanford Co.7; Parker, HW-24131, 7; Parker, HW-25994, 6.

37. H. M. Parker, "Columbia River Situation—A Semi-Technical Review," HW-32809, GE Hanford Co., August 19, 1954, 1–4.

38. C. L. Dunham, "Information on Columbia River Situation," Enclosure C (October 26, 1954) of "Columbia River Contamination" 30598, U.S. AEC, November 22, 1954; J. C. Bugher, "Columbia River Contamination," Enclosure D (November 17, 1954) of "Columbia River Contamination" 30598, U.S. AEC, November 22, 1954.

39. DeNeal, DUN-6888, 12–13.

40. Davis, Watson, and Palmiter, HW-36074, 12.

41. Ibid., 52, 62, 65.

42. H. A. Kornberg, "The Absorption and Toxicity of Several Radioactive Substances in Plants and Animals," CIC-26514, GE Hanford Co., 1955, 2, 3, 7.

43. Ibid., 16.

44. R. McMurray, "Columbia River Aspects of Increased Production," HW-41049, GE Hanford Co., January 25, 1956, 2–3.

45. Watson and Davis, HW-48523, 29–30.

46. B. V. Andersen and J. K. Soldat, "The Regional Monitoring Program," HW-52038, GE Hanford Co., August 26, 1957, 10, 5.

47. R. B. Hall, "Radioactivity in 100-Area Water," HW-52410, GE Hanford Co., September 9, 1957, 1, 3.

48. DeNeal, DUN-6888, 18.

49. B. V. Andersen, "Regional Monitoring Activities, November 1958," HW-58330, GE Hanford Co., December 4, 1958, 4.

50. J. W. Healy, "Quarterly Progress Report, Research and Development Activities for October–December 1958," HW-58833, GE Hanford Co., January 12, 1959, 14.

51. DeNeal, DUN-6888, 49.

52. R. B. Hall, "Waste Disposal Criteria, Existing Reactor Expansion Study," HW-65733, GE Hanford Co., November 17, 1959, 1, 3–4.

53. Hall and Jerman, HW-63653, 12.

54. Ibid., 6–8.

55. Foster and Junkins, HW-63654, 27–28, 37; See also: R. L. Junkins, E. C. Watson, I. C. Nelson, and R. C. Henle, "Evaluation of Radiological Conditions in the Vicinity of Hanford for 1959," HW-64371, GE Hanford Co., May 9, 1960, 20.

56. DeNeal, DUN-6888, 49. Note: The reactors reached these levels again in 1964.

57. Hanford Laboratories Operation and Irradiation Processing Department, HW-70562,3; J. K. Soldat, "A Compilation of Basic Data Relating to the Columbia River, Section 8, Dispersion of Reactor Effluent in the Columbia River," HW-69369, GE Hanford Co., November 16, 1962.

58. DeNeal, DUN-6888, 21, 51; See also M. H. Schack, "Effluent System Capacity, Increased Process Water Flow, 100-K, Project CGI-883," HW-67444, GE Hanford Co., November 18, 1960.

59. DeNeal, DUN-6888, 22–23.

60. R. F. Foster, "Report to the Working Committee for Columbia River Studies on Progress Since September 1962 for Projects Carried Out by the General Electric Company at Hanford," HW-80649, GE Hanford Co., February 1964, 27.

61. U.S. Department of Energy, "N Reactor Comprehensive Treatment Report, Hanford Site, Washington," DOE/RL-96-91, Vol. I, U.S. Department of Energy (Richland, WA), September 1997, III-1.

62. "Hanford to Cut Back in 1965," in *Tri-City Herald* (Kennewick, WA), January 8, 1964, A1, citing L. B. Johnson.

63. DeNeal, DUN-6888, 26.

64. R. F. Foster, and R. H. Wilson, "Evaluation of Radiological Conditions in the Vicinity of Hanford for 1964," BNWL-90, Battelle Northwest Laboratories (BNWL) (Richland, WA), July 1965, Figure E, 18.

65. DeNeal, DUN-6888, 27–29.

66. R. G. Geier, "Monthly Report, Contamination Control—Columbia River, December 1965," DUN-460, DUN, January 3, 1966, 5–6.

67. Ibid., 10. See also W. K. Hensley, and L. A. Rogers, "Plutonium Recycle Test Reactor (PRTR) Accident," NUREG/CR-3669, PNL-5003, Pacific Northwest Laboratory (PNL) (Richland, WA), April 1984.

68. DeNeal, DUN-6888, 30–33.

69. D. G. Watson, C. E. Cushing, C. C. Coutant, and W. L. Templeton, "Radioecological Studies on the Columbia River," BNWL-1377, BNWL, May 1970, 29–32.

70. DeNeal, DUN-6888, 34–37.

71. J. P. Corley and C. B. Wooldridge, "Evaluation of Radiological Conditions in the Vicinity of Hanford for 1967," BNWL-983, BNWL, March 1969, 23.

72. DeNeal, DUN-6888, 38–41.

73. J. W. Ballowe, "Project Management Plan, Project 78-18-j: N Reactor Environmental Improvements, Richland, Washington," UNI-829, UNI, July 29, 1977; Hopkins, UNI-M-94.

74. C. E. Cushing, D. G. Watson, A. J. Scott, and J.M. Gurtison, "Decline of Radionuclides in Columbia River Biota," PNL-3269, PNL, March 1980, Abstract 1, 19.

75. J. J. Fix, "Association of Long-Lived Radioactivity with Sediment Along the Columbia River Shoreline, Islands, Bottom and Slough Area," BNWL-SA-5484, BNWL, April 1976, 2. See also: D. E. Robertson and J. J. Fix, "Association of Hanford Origin Radionuclides with Columbia River Sediment," BNWL-2305, BNWL, August 1977: U.S. Energy Research and Development Administration, ERDA-1538, Vo. I, I. 3–64.

76. Gerber, WHC-EN-RPT-004, 13; DuPont, HAN-73214, Book 15, 115–116; B.J. Borgmier, "Bibliography on Slug Failures Occurring in the Hanford Production Reactors," HW-21199, GE Hanford Co., May 24, 1951.

77. DeNeal, DUN-6888, 44; Gross, HW-8660, 1.

78. DeNeal, DUN-6888, 44.

79. Parker, HW-24356, 5.

80. DeNeal, DUN-6888, 44.

81. Ibid., 44; J.D. McCormack and L.C. Schwendiman, "Significance of Rupture Debris in the Columbia River," HW-61325, GE Hanford Co., August 17, 1959, 3–4, 12. Note: Strontium 89 and strontium 90 were the strontium isotopes of concern.

82. DeNeal, DUN-6888, 44.

83. Department of Energy, DOE/AD-0015, 2-3.

84. Ibid., 2-4.

85. G. E. Backman, "Summary of Environmental Contamination Incidents at Hanford, 1958–1964," HW-84619, GE Hanford Co., April 12, 1965, 4; Department of Energy, DOE/AD-0015, 2-30, 2-40.

86. Department of Energy, DOE/AD-0015, 2-29, 2-32, 2-25, 2-26, 2-75.

87. Ibid., 2-121.

88. Ibid., 2-38.

89. Ibid., 2-42.

90. Ibid., 2-27, 2-20.

91. Foster, Davis, and Olson, HW-33366, 3.

92. Gerber, WHC-EN-RPT-004, 1-2.

93. Olson, HW-11642, 2.

94. Foster, HW-23793, 3.

95. R. B. Richman, "Water Treatment History at Hanford Reactors Through August 20, 1956," HW-45070, GE Hanford Co., August 21, 1956, 2–5.

96. Parker, HW-32809, 1, 3.

97. Hanford Laboratories Operation and Irradiation Processing Department, "Monthly Report, Contamination Control—Columbia River," HW-70526, GE Hanford Co., July 21, 1961, 4.

98. Foster, Davis, and Olson, HW-33366, 4, 11, 13, 65.

99. Mickelson, HW-39558, 15.

100. R. F. Foster, "Recommended Limit on Addition of Dichromate to the Columbia River," HW-49713, GE Hanford Co., April 17, 1957, 2–3, 5–6.

101. J. K. Soldat, "Chemical Characteristics: Sections 4.5 and 4.6," in Foster, HW-69368, 4.6.0-1–4.6.1.4.

102. Ibid., 4.6.4-1.

103. DuPont, HAN-73214, Book 11, 76–77, 110; C.P. Kidder, "100-B Unit Purge—Part I—Pressure Drop Studies," HW-3-2224, Hanford Engineer Works, April 16, 1945; W.R. Conley, "Production Test 105-3-MR: The Use of Dicalite Diatomaceous Earth as a Purge Material in the 100 Areas," HW-24055, GE Hanford Co., April 8, 1952. Note: Dicalite is a registered trademarked product of the Great Lakes Carbon Corp. of New York, NY.

104. J. W. Healy, "The Effect of Purging During Pile Operation on the Effluent Water," HW-24578, GE Hanford Co., May 27, 1952.

105. W. Y. Matsumoto, "The Effect of Reactor Purges on Radioactive Contamination of Coolant," HW-43830, GE Hanford Co., June 20, 1956.

106. D. J. Brown, "Chemical Effluents Technology Waste Disposal Investigations, April, May, June 1957," HW-53225, GE Hanford Co., September 27, 1957, 9–10.

107. W. N. Koop, "Effect of Reactor Purges When River Temperature Exceeds 15C," HW-50601, GE Hanford Co., June 10, 1957, 2.

108. DeNeal, DUN-6888, 6–7, 49; Hall and Jerman, HW-64517, 2.

109. Healy, HW-3442.

110. Parker, HW-22576, 7–8.

111. Parker, HW-24356, 6.

112. Parker, HW-32809, 1, 3.

113. Foster, Davis, and Olson, HW-33366, 4, 13.

114. McMurray, HW-41049, 2–3.

115. J. W. Healy, "Quarterly Progress Report, Research and Development Activities for October—December 1956," HW-48044, GE Hanford Co., January 28, 1957, 9.

116. J. W. Healy, "Quarterly Progress Report, Research and Development Activities for October—December 1957," HW-54938, GE Hanford Co., February 12, 1958, 5.

117. R. F. Foster, "The Effect on Fish of Increasing the Temperature of the Columbia River," HW-54858, GE Hanford Co., March 14, 1958, 16, 18.

118. Hall, HW-65733, 3,7.

119. H. A. Kramer, "Artificial Cooling of the Columbia River by Dam Regulation, 1959," HW-65767, GE Hanford Co., June 20, 1960, 4–5, 17.

120. H. A. Kramer, "Artificial Cooling of the Columbia River by Dam Regulation, 1960," HW-68337, GE Hanford Co., February 15, 1967; H. A. Kramer, and J.P. Corley, "Use of Lake Roosevelt Storage to Lower River Temperatures," HW-SA-2301, GE Hanford Co., October 26, 1961; H. A. Kramer, "Artificial Cooling of the Columbia River by Dam Regulation—1962," HW-76887 ADD, GE Hanford Co., 1963, 3; R. T. Jaske, "An Evaluation of the Use of Selective Discharges from Lake Roosevelt to Cool the Columbia River," BNWL-208, BNWL, February 1966.

121. J. W. Ballowe, "Summary Report, Alternate Methods of Reactor Treatment and Disposal," DUN-2231, DUN., March 22, 1967, 11

122. J. W. Ballowe and R. G. Geier, "Summary Report—Reduction of Radioactivity and Thermal Energy Discharged to the Columbia River," DUN-2273, DUN, March 31, 1967, 4, 15.

123. J. F. Honstead, "Disposal of Reactor Effluent Through an Inland Lake System," HW-39465, GE Hanford Co., October 11, 1955.

124. Mickelson, HW-39558, 5.

125. J. J. Fuquay, "A Preliminary Appraisal of Meteorological Aspects of Disposal of Reactor Effluent Through an Inland Lake System," HW-40371, GE Hanford Co., December 8, 1955.

126. Keene, HW-72819, 3, 5.

127. Ibid., 5, 2.

128. Ibid., 2–3, 8–9.

129. J. W. Ballowe, "Summary Report, Alternate Methods of Reactor Effluent Treatment and Disposal," DUN-2231, DUN, March 22, 1967; R. G. Geier, "Quarterly Report—Contamination Control—Columbia River—October—December 1967," DUN-3935, DUN, March 11, 1968.

130. Ballowe and Geier, DUN-2273, 4–5, 13–15.

131. Ibid., 4, 16.

132. Healy, HW-60529, 3, 6, 8.

133. Silker, HW-56366, 3, 6.

134. Silker, HW-59029, 3–4.

135. H.G. Rieck, "Pilot Scale Aluminum Bed Decontamination of Reactor Effluent, Final Report," HW-72215, GE Hanford Co., March 1962.

136. Hall, HW-68521, 1, 6–7.

137. Hanford Laboratories Operation and Irradiation Processing Department, HW-70526, 4–5; R. G. Geier, and F.W. VanWormer, "Water Treatment Program—Old Reactors," HW-75609, GE Hanford Co., November 19, 1962, 3–5; R. G. Geier, "December Monthly Report, Contamination Control—Columbia River," HW-75949, GE Hanford Co., December 20, 1962, 4–5.

138. L. D. Perrigo, "Status Report: Production Reactor Process Tube Film Composition and Radionuclide Inventory Studies," HW-76557, GE Hanford Co., February 12, 1963, 2–5; R. G. Geier, Monthly Report, Contamination Control—Columbia River, December 1963," HW-80095, GE Hanford Co., December 20, 1963, 4–5.

139. R. G. Geier, "Monthly Report, Contamination Control—Columbia River, December 1965," DUN-460, DUN, January 3, 1966; R. G. Geier, "Quarterly Report, Contamination Control—Columbia River, October—December 1967," DUN-3935, DUN, March 11, 1968, 4–5.

140. D. E. Robertson and R. W. Perkins, "Reduction of Radionuclides in Reactor Effluent Water: Final Report on the Effect of Chemical Additives and Coating Materials on the Adsorption of Radionuclide Parent Elements in the Process Water on Aluminum Surfaces," HW-80557, GE Hanford Co., December 23, 1963.

141. W. B. Silker and C. W. Thomas, "Effect of Deionized Water on Reactor Effluent Activities, Part I: New Tubes and New Fuel Changes, HW-81412, GE Hanford Co., October 1964.

142. Gamertsfelder, HW-7-5934, 5.

143. R. C. Walker, E. L. Etheridge, and D. F. Watson, "Conversion of a 'K' Reactor to Single-Purpose Power Recovery—Feasibility Study," HW-55910, GE Hanford Co., May 26, 1958.

144. Ballowe, DUN-2231, 9; Ballowe and Geier, DUN-2273, 4, 15. Note: This amount was more than half of the original cost of each K Reactor.

145. Ballowe and Geier, DUN-2273, 17.

LOOKING BACK, LOOKING FORWARD

CHAPTER TEN

Atomicalia
Collecting and Exhibiting
Manhattan Project Material Culture

Mick Broderick

Radioactive monsters, utopian atom-powered cities, exploding plan-
ets, weird ray devices, and many other images have crept into the way
everyone thinks about nuclear energy, whether that energy is used in
weapons or in civilian reactors.

Spencer Weart, *Nuclear Fear*, 1988

The atomic bomb that leveled Hiroshima also blasted openings into a
netherworld of consciousness where victory and defeat, enemy and self,
threatened to merge [...] children of the 1950s grasped the pleasures of
victory culture as an act of faith, and the horrors of nuclear culture as
an act of faithless mockery, and held both the triumph and the mock-
ing horror close without necessarily experiencing them as contraries.
In this way, they caught the essence of the adult culture at that time,
which—despite America's dominant economic and military position in
the world—was not one of triumph, but of triumphalist despair.

Tom Engelhardt, *The End of Victory Culture*, 1995

With these observations in mind, this essay explores an ongoing
material culture project undertaken in the lead-up to the 50th
anniversary of the development, testing, and use of atomic bombs at
Hiroshima and Nagasaki. I drew principally from my own personal collection of atomic artifacts in the design and curation of an exhibition for

the state museum of Western Australia, entitled *Half Lives: Experiencing the Atomic Age*, which ran from December 2004 to February 2005, and was subsequently repurposed as *Atomicalia* in Japan and Canada (2009-2014.) These exhibitions showcased displays of transnational material culture objects produced prior to, during, and beyond the Cold War and how the items reflect the nuclear *episteme,* a term Michel Foucault deployed to denote a period's epistemology, its governance of knowledge and power.[1] For this reason I do not distinguish between atomic energy used for weapons or militaristic purposes, or civilian nuclear power and the associated nuclear fuel cycle.

It is not my intention to celebrate the material I am presenting here—although as a baby-boomer the materials resonate strongly with me. I have collected such items for over two decades, and researched how they are advertised, promoted, and sold online, chiefly to demonstrate the lasting effects of how ideas about nuclear energy and atomic weapons remain circulating within our culture, outside of any obvious political context. One of my aims is to draw attention to international exhibition audiences how (mostly Western) consumers embrace the concepts and the branding of atomic weapons and nuclear energy in complex and sometimes contradictory ways.[2] The process of purchasing goods is often considered an unconscious or impulsive act—operating at a deeply symbolic level, where the ideology surrounding these (atomic) products is either unrecognized or accepted as natural and normative so as not to be overtly questioned.[3] Hence, the imagery accompanying this essay, and the language that online sellers of *atomicalia* deploy to promote their goods, remains an underexplored window into the nuclear era and the Manhattan Project that propelled it.

ATOMICALIA AND (AMERICAN) CULTURE

One of the principal organizing strategies of this and subsequent exhibitions—and my ongoing approach to collecting—is to highlight the mundane, the banal, the kitsch, and the everyday alongside conventional historical and commemorative items. Hence, the term *atomicalia* can be considered a taxonomy that arranges material under the broad rubric of "Atomic Age artifacts." A significant subset of *atomicalia* is the collection and recirculation of Manhattan Project artifacts and the spin-off products relating to the development and use of the atomic weapons. *Atomicalia* is deliberately eclectic and non-exclusionary. It seeks to discover how pervasive and successful nuclear images, iconography, themes, and identities have been at entering our daily lives, often without conscious recognition. While *atomicalia* frequently operates at a symbolic level, both its overt and hidden identifications paradoxically work in unison—and in opposition. This potential binary antagonism, its negotiation and the extensive gradation between ends of the "pro" or "anti" nuclear spectrum, has been described by other scholars of the atomic age. As the opening epigrams suggest, Spencer Weart has demonstrated in *Nuclear Fear* that we have the capacity to entertain nuclear imagery in this dualistic mode.[4] Similarly, Tom Engelhardt in *The End of Victory Culture* highlights the seemingly contradictory play and associations held by baby-boomer children throughout the Cold War.[5]

Informing the collection strategy around this cultural detritus are several key theoretical assumptions. As Raymond Williams contended, culture is "ordinary" and that should be the foundational premise from which any such approach must start.[6] Adopting a perspective of the "culture of everyday life" eschews the conventional, narrow view of nuclear energy politics—frequently dichotomized into either a "for" or "against" debate. Importantly, by foregrounding the *ordinariness* of culture, the Zeitgeist of the atomic age can be evoked. As Norbert Elias advocated, a nonpartisan approach to daily interactions in leisure and lifestyle is the initial key to understanding our social responses to many issues.[7]

The politicized arguments surrounding nuclearism have for too long ignored the deep cultural penetration that nuclear images and iconography have had on western society. Michel de Certeau's cultural methodology shows us, albeit in another context, the need to contest notions of "passivity" in consumption.[8] His work undermines the essentializing, hypodermic model of Frankfurt School theorists of media and cultural transmission, where messages are supposedly "injected" into us from outside unproblematically. Hence, I argue that the cultural manifestations of our nuclear age, nascent in the pre-Manhattan Project era, can be read "textually" not just in the abundant programming of film, television, radio, and advertising, but equally through the material display of atomic themes and tropes adopted by game and toy makers and in decor, apparel, cosmetics, food stuffs, and household appliances, among other everyday applications.

By emphasizing that "modernity and everydayness constitute a deep structure that critical analysis can uncover," Henri Lefebvre noted that such enquiry need not be intrinsically "banal" in the pejorative sense.[9] My exhibitions of *atomicalia*, for example, explore how the nuclear era has infused our individual subjectivity and self-expression, throughout the Cold War and beyond, ranging from the customized motor vehicle license plates we purchase, through to the postage stamps we choose to adorn mail (see figures 10.8 and 10.15). From the disposable packaging of merchandise, to the legal tender of national currency, the atom is ubiquitous. *Atomicalia* shows how both industrial design and promotional branding have employed nuclear motifs. It provokes consideration as to how such imagery can have very different connotative (implied) and denotative (literal, or primary) meanings depending upon the artifacts' utility, as well as their national and historical contexts.

Today there is an abundant historical archive of such material, mostly found and traded online, as well as the continuous cultural and economic production of such goods. Importantly, nuclear collectables do not end with the Cold War. They are as abundant and prolific today as they were in the 1950s through the 1980s. A recent eBay keyword description search on "atomic" found over 250,000 hits, and for "nuclear" over 180,000 items.[10] *Atomicalia* reveals the diversity of cultural product available for mass consumption which spans the multiplicity of media: newspapers; book and magazine covers; comics; advertising; radio and television programming; cinema, DVD, and online downloads; arcade, computer, and hand-held digital games; and LPs, 45s, CDs, and digital sound files. Hence, the nuclear iconography of this *episteme* is abundant—one needs only to consider the awe-inspiring mushroom cloud, the ubiquitous symbol of the atom with its orbiting outer particles and inner nucleus, or the radiation warning trefoil. Even the branded "CD" of Civil Defense maintains a cultural cache amongst collectors.

But what are the inherent and latent semiotics of these items as cultural talismans of the legacy of the Manhattan Project? What factors determine the differing connotations that arise from proximity to these products at different times and in different national contexts? What does it mean, for example, to wash your clothes in Atomic Soap Suds, to apply Cramer's Atomic Balm—even though it cautions: "This product is intended to and will cause mild skin irritation"? To shave using Atomic safety razors or a Ronson electric razor adorned with a spiraling atomic nucleus, and apply nuclear Graphite antiperspirant in the form of a missile? To use A-bomb shaped salt and pepper shakers or cook with an Atom Pop Corn popper? To drink Atomic Nitro energy juice, or mix an Atomic Cocktail? To eat candies like Nuclear Bon-Bons, Atomic Fireballs, Wally Warheads or Nestlé's Nuclear chocolate? To consume Samboy Atomic tomato crisps or drink coffee from a Milan Atomic espresso machine, now a highly prized "retro" *objet d'art*? To light your cigarettes with a "3 Mile Island" or radiation hazard adorned Zippo, filled with Atomic Lighter Fluid, and stub your butt out in a casino ashtray promoting the winning jackpot as "fallout" from Atomic poker machines, or other commemorative ashtrays promoting civilian nuclear power plants? What does it mean to wear Atomic Perfume dispensed from an A-bomb shaped cologne bottle, or to don cufflinks displaying miniature cooling towers with matching tie pin, or a wristwatch featuring the nuclear trefoil, a belt buckle shaped as a nuclear utility with reactor, or boxer shorts with radiation hazard symbols warning of "restricted entry"? To display sterling silver earrings fashioned after the Little Boy and Fat Man atomic bombs that decimated Hiroshima and Nagasaki, or to wear irradiated uranium glass earrings that glow under UV light? Such items (figures 10.1–10.15) are the veritable tip of the *atomicalia* iceberg and indicate that we have truly learned, if not to love atomic energy, then at least to consume it metaphorically in our daily lives and across all facets of our sociocultural *habitus*.[11]

Traditionally, such questions of consumption in cultural studies have adopted post-Marxist ideological analysis but following from Ben High-more's scholarship *atomicalia* can be open to the suggestion that everyday culture is characterized by "ambiguities, instabilities and equivocation" while providing "a site of resistance, revolution, and transformation."[12] Despite an object's deliberate atomic design and commodification, there is no guarantee that it will be interpreted or consumed in the manner intended by the manufacturer or seller, especially between generations and across nations.

A common assumption made by scholars, even those engaged in "nuclear criticism," is that the Atomic Age is long over.[13] However, the

10.1. Cramer Atomic Balm. *Author's collection.*

10.2. Atomic Fire Ball candy. *Author's collection.*

10.3. Samboy Atomic Tomato crisps. *Author's collection.*

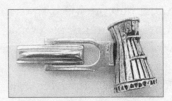

10.4. Cuff link in the shape of a nuclear reactor. *Author's collection.*

10.5. Graphite brand deodorant with radiation symbol. *Author's collection.*

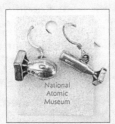

10.6. Earrings in the shape of atomic bombs. *Author's collection.*

10.7. Ronson electric razor with atomic symbol. *Author's collection.*

10.8. State license plates with nuclear themes. *Author's collection.*

10.9. Atomic brand instant lighter fluid. *Author's collection.*

10.10. Zippo brand lighter with radiation symbol. *Author's collection.*

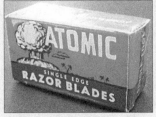

10.12. Atomic brand razor blades. *Author's collection.*

10.13. Atomic Soap "disintegrates dirt." *Author's collection.*

10.11. Atom Bomb perfume in bomb-shaped bottle. *Author's collection.*

10.14. Bomb-shaped salt and pepper shakers. *Author's collection.*

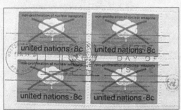

10.15. United Nations postage stamps with nuclear weapon theme. *Author's collection.*

ongoing production of *atomicalia* and the audiences attracted to the *Half Lives* and *Atomicalia* exhibitions in Australia, Japan, and Canada, attests that atomic/nuclear popular culture is as strong as it has ever been, despite the frequently proclaimed "end of the Cold War" in 1991.

Most experts engaging with popular cultural analysis of the period have concentrated on the mass media, particularly film, television, and literature, with only limited discussion of artifacts and iconography.[14] Despite the close media textual readings—and the application of political, historical, and psychological methodologies to investigating the atomic age—few of these studies have considered these abundant material popular culture items as worthy of serious attention. Only a handful of scholarly works have engaged with the material manifestation of the atomic era. Paul Boyer, for instance, suggests that the materials produced by toy manufacturers reflected popular concerns or naivety of the period, whereas the post-Cold War era often promotes cynical associations.[15] Michael Amundson is more revealing of the nuances associated with nuclear age artifacts when discussing the uranium "craze" of the early 1950s but treats this episode nostalgically and hermetically contained within the Cold War period.[16] In a provocative essay, Dina Titus considers the iconography of the mushroom cloud in terms of kitsch, but nevertheless suggests the deployment of "positive" atomic images by government, military, and industry to promote nuclear technology as being absorbed uncritically and homogenously by the intended audience.[17] It is often assumed that we, as consumer-citizens, are not sophisticated enough to understand the threat, danger, or expediency of such sales propaganda.

Few if any of these commentators consider why material objects signifying nuclear technologies and related themes—outside of the conventionally defined atomic age or Cold War—continue to be produced, have currency, and remain highly popular internationally, well into the 21st century.

CASE STUDIES

The following *atomicalia* case studies are of particular relevance to the Manhattan Project and the use of atom bombs at the close of the Pacific war in 1945. As numerous historians have noted (e.g. Gar Alperovitz and Barton Bernstein), in many ways the atom bombings heralded the start of the Cold War. More than seven decades on, the continued circulation and consumption of Manhattan Project and Hiroshima or Nagasaki collectables into the post-Cold War era signifies a conflicting mix of pride, nostalgia, denial, guilt, and potential fears of national vulnerability.

The exponential rise of items searched for, traded, and collected on websites such as eBay is now a technological and cultural mode of daily life that we take for granted. This globalized online commerce enables access to material previously the domain of auction houses or the labor-intensive trawling through of newspaper classified advertising or visiting antique stores or neighborhood yard and garage sales. This new mode of international commodity exchange suggests evolving cultural practices beyond those identified by 1980s cultural theorists concerned with everyday "shopping."[18] *Atomicalia* and its exhibition not only showcases those objects under scrutiny, but also demonstrates their changing patterns of usage, whereby the goods are no longer necessarily sought for their original design, function, and purpose, but have now been invested with an "aura" as a collectable of the Atomic Age, connoting something over and above their prior commercial utility.

Each of the following case study objects has been previously posted online on eBay as available for purchase from anywhere in the world. The composition here recreates the look of the original website design, typography, and layout while maintaining any errors in spelling and grammar in order to replicate the original pages. Entries have been edited to ensure anonymity of the online vendors. This small sample of artifacts, directly relating to the bombing of Hiroshima and Nagasaki, originates from online sellers in the United States, and the items are mostly pitched to a domestic American audience. Rather than simply allowing the artifacts to represent themselves, most vendors employ discourse to "sell" the items, especially to heighten the importance of the products' "authenticity" or rarity in order to maximize their market values.

But what can we make of the explicit sales rhetoric invoking patriotism, heroism, sacrifice, and, in some instances, jingoism, racism, and xenophobia? As Engelhardt asserts, "the moment of total victory" in World War II came with the dropping of the atom bombs, but it also, paradoxically, created a perpetual sense of national crisis: "How Americans have dealt (or failed to deal) with the implications of global dominance to which their history had brought them in 1945, and how they have (or have not) come to terms with the slow-motion collapse of a heroic war ethos thereafter, are central themes underlying American popular culture from 1945 on."[19]

Enola Gay autographed 8x10 (City in Ruins)
Currently $75.00

Description - This is superb picture of the after effects of the A-BOMB exploding. It is signed by 4 members of the ENOLA GAY flight (Colonel Lieutenant Paul W. Tibbets [pilot], Major Thomas W. Ferebee [bombardier] and Captain Theordore "Dutch" Van Kirk [navigator], Pfc Richard H. Nelson [radio operator]. A great chance to get a piece of history. We have sold lots of Enola Gay Pictures by the 3 and they sell for $100+ each. This is the first time we have offered them signed by the 4. Richard Nelson is a very very tough autograph. Autographs were obtained IN PERSON at a private signing Memorial Day Weekend 1999 in Branson Mo. Once they are gone they are gone. We will sell this at NO RESERVE to the high bidder. S/H is $5.00

All items come w/COA from INSIDE-EDGE. Check our feedback most users feel we are one of the FASTEST shippers on eBay. We are Powerseller Gold Members as well been listed in eBay's TOP 100!!!! Bid with confidence we have 7000+ POSITIVE Comments. Please email us if images dont load or are incorrect.

As a World War II celebrity, *Enola Gay* pilot Paul Tibbets was somewhat notorious for signing almost anything in order to receive a small fee. His signature adorns magazines, books, photos, model aircraft kits, movie posters, checks, currency, commemorative stamps, and first-day-of-issue postage stamp covers. Ironically, his ubiquitous imprimatur evident on these online goods and his willingness to autograph any merchandise actually served to diminish the value for buyer-investors, rather than increase the item's worth. My own *atomicalia* collection contains several Tibbets' signed artifacts, including a 1946 first edition hardcover of John Hersey's *Hiroshima*; an autographed one-sheet movie poster of

MGM's biopic of Tibbets, *Above and Beyond* (1952); and a faux philatelic first day of non-issue cover featuring the withdrawn 1995 U.S. postage stamp, "Atomic bombs end WWII." On occasion Tibbets must have been oblivious, inattentive, or ignorant of the objects placed before him. For example, over the past 20 years, the autograph "Paul Tibbets, Pilot" has accompanied several National Archives photographs circulating on eBay depicting the distinctive *Nagasaki* mushroom cloud (!), erroneously captioned "HIROSHIMA, JAPAN August 6, 1945."

HIROSHIMA, JAPAN   August 6, 1945

In a similar vein, consider how the following item draws upon patriotism and hero mythology to reinforce its pitch at the end of the description. There is no discussion about the question of profiteering from these unique atomic bomb artifacts, something the U.S. government unsuccessfully challenged in court on several counts in order to prevent these unique items being purchased by private parties.[20]

EXCLUSIVE ~ ATOMIC BOMB (HIROSHIMA) "LITTLE-BOY" PLUGS
Signed by: Enola Gay's Weapon Officer ~ Morris Jeppson
Starting bid: **US $39.9**

Enola Gay" & "Little Boy" ~ Auction ~ June 11,2002
Morris "Dick" Jeppson *(Weapon Officer)* & "Dutch" Van Kirk *(Navigator)*

Hiroshima Bomb "Little Boy" Arming Plugs
Property of Morris "Dick" Jeppson

Arming & Safety Plugs ~ Only Surviving Artifacts
From the 1ˢᵗ Atomic Bomb Used in Warfare
(These Plugs are from the Little Boy Bomb)

EXCLUSIVE ~ I am the ONLY seller selling this item (designed by me and signed by Jeppson - In Person - in 2005) ~ This display item (color photograph signed by Jeppson) shows the - ACTUAL ARMING PLUGS - that were attached - by Morris Jeppson - on the Hiroshima Atomic Bomb "Little Boy" on August 6th, 1945. NOTE: These are NOT the actual $150,000 plugs - but an autographed 8x10 color display.

The plugs were offered for auction with Butterfield & Butterfield on June 11, 2002 by Morris Jeppson ~ The Actual 1945 Plugs did sale at auction for $150,000 ~ This is a color copy of the original auction page from Butterfield & Butterfield with a HANDSIGNED, AUTHENTIC, IN-PERSON, ORIGINAL AUTOGRAPH by Morris Jeppson (weapon officer on the Enola Gay).

Approx size 8.5" x 11" - as shown - Signature is very sharp - Would frame up nicely! To end WWII, in August 1945 (6th & 9th) two Atomic Bombs were dropped over Japan (Hiroshima & Nagasaki). The 1st was dropped by "The Enola Gay" with 12 crewmembers on board. The 2nd was dropped by "The BocksCar" with a crew of 13 on board.
Right now there are 4 TOTAL (out of the 25 crewmembers) still alive. (3 Enola Gay & 1 Bocks Car crewmembers)
I am THE foremost collector & seller of Atomic Bomb Missions authentically autographed item on eBay. I also possess one of the largest collections of authentically autographed items by the Atomic Bomb Crewmembers, Scientists & Coordinators in the U.S. (500+ different items). All these men are my "heroes of August 1945" - bar none!

The next object is described with a feigned naivety to lure the buyer. It appeals, somewhat disingenuously, to "someone who is into these war history type of collectables." The rhetoric used here is almost one of embarrassment or disdain, not from selling the item, but directed towards the "type" of person who would want to buy it. One can only wonder what Americans would think about a similar listing appearing on eBay auctioning off rubble from the 1941 Pearl Harbor bombing to Japanese "war history" buffs?

ATOMIC HIROSHIMA WWII PC./BUILDING *JAPANESE WRITING*

Current bid: **US $102.50**
Ships to: Worldwide
Item location: Northeast P.A., United States
History: 3 bids

Description:

FOR YOUR BIDDING CONSIDERATIONS IN THIS AUCTION, WE BRING TO YOU WHAT WE BELEIVE TO BE IS AN ORIGINAL PIECE OF A BUILDING FROM THE HIROSHIMA ATOMIC BLAST FROM WWII. THIS PIECE OF CONCRETE WAS FROM A FRIEND OF OURS MOTHERS HOUSE THAT WE HELPED CLEAN OUT AFTER SHE PASSED ON. WE CAN NOT READ JAPANESE AND HAVE NOT FOUND ANY ONE IN OUR AREA THAT DOES, SO WE CAN'T DECIPHER THE WRITING ON IT FOR YOU.

THERE IS NOT MUCH ELSE TO DESCRIBE, AS THE PICTURES PRETTY MUCH SAY IT ALL. WE FEEL THIS IS A TERRIFIC OPPORTUNITY FOR SOMEONE WHO IS INTO THESE WAR HISTORY TYPE OF COLLECTIBLES TO ADD TO THIER COLLECTION.

PLEASE ASK ANY QUESTIONS AT ALL THAT YOU MIGHT HAVE BEFORE BIDDING, AND WE WILL ANSWER THEM IN A TIMELY AND HONEST MANNER, THANK YOU.

**IF BY ANY CHANCE SOMEONE OUT THERE SEES THIS AND CAN DECIPHER THE WRITING, WOULD YOU PLEASE BE KIND ENOUGH TO EMAIL AND TELL US WHAT THIS WRITING MEANS OR SAYS, THANK YOU.

The following item also invokes American patriotism in its advertising pitch, but what is to be made of the prominent display of double Confederate civil war flags which appeared to flutter in the online listing, decidedly at odds with the purported phrase "home of the free." Free for whom? The implication of this item description is that the sand fused into glass at the Trinity site in New Mexico, and that of the apparent randomly collected new soil from Nagasaki, are equivalent to "other significant battle areas around the world." However, neither of these sites experienced "battle"—both faced the catastrophic and annihilating heat, blast, and radiation from a single atomic detonation, contaminating the area while irradiating American downwinders and Japanese *hibakusha*.

1760-1st Atomic Bomb Relics - WOW!!!
Starting <u>bid:US</u> **$24.95**

Deep in the heart of beautiful East Texas

<u>**The home of the free because of the brave.**</u>

This display framed display contains a piece of <u>Trinitite</u> on the left and on the right is a small amount of earth from near ground zero at Nagasaki, Japan where Fat Man was dropped on August 9, 1945 - just 24 days after the first test. The background of the display contains some information about the subject, a picture of Fat Man and a picture of Nagasaki after the blast. It also comes with my <u>COA which</u> has more details. The COA reads as follows: "This display is one of a series of 32 unnumbered displays I created. <u>The Nagasaki soil was collected by retired US Marine Sgt. Chuck Truitt</u> when he visited Nagasaki. Chuck collects sand from beaches and soil from other significant battle areas around the world. He was a missionary in Okinawa for a number of years after his medical retirement and as of March 2004 he merged his Okinawa mission with another, took leave to return to the US and was looking for another mission.

The next object is essentially self-explanatory, and perhaps the most obvious example conforming to Englehardt's thesis concerning post-war American children entertaining—and by implication, haunting—themselves with toys and games about atomic war.

Post WWII ATOMIC BOMB Game Toy Hiroshima, Nagasaki

Starting bid: **US $39.99**
Item location: Green Bay, Wisconsin, United States

Description: This auction is for an Atomic Bomb game by A. C. Gilbert Co. from late 1945. This is a game of skill to try and get the miniature bombs to land in the two recesses on the Japanese cities of Nagasaki and Hiroshima by moving the box. The box is a cardboard frame holding a piece of glass covering the game. The game measures 1 x 3 1/4 x 4 1/4 inches. The game does show use. Take a look at the other WWII era items we have for sale to add to your collection! The winning bidder will pay postage and handling using USPS First Class Priority Mail. If you have any questions,
Please
email me.
Thanks!

In the following online advertisement, the rhetoric belies anxiety or unease in which the vendor demonstrates antipathy towards the item up for auction. The T-shirt is presented as a satirical rebuff against the perception during the late Cold War period that America had lost its global market dominance to Japan, as "Number One." Note how the vendor repeatedly—three times within three lines—informs the readers how "funny" the T-shirt is. This emphasis appears as a self-conscious attempt to over-compensate for an unmistakably xenophobic overtone.

Vintage Japan atom Bomb shirt punk <u>emo</u> indie

Vintage Japan Bomb shirt

Currently: US $6.00

This is a cool vintage shirt. It is an off white shirt with a blue and hot pink print. The picture of the shirt is of a mushroom cloud like that left by an atomic bomb. The mushroom cloud is in blue. Across the chest of the shirt, in blue block letters it says: MADE IN AMERICA.Then in the top part of the cloud, in pink it says: BY LAZY, UNPRODUCTIVE AMERICAN WORKERS And below the cloud, in pink block letters it says: TESTED IN JAPAN! This shirt is cool and funny. The size is Large. The shirt is 50/50 cotton/polyester. The shirt brand is DELTA, made in the U.S.A. This shirt is awesome and funny. BID WITH CONFIDENCE. No holes, rips, or stains to worry about! This shirt looks good and fits great. It has a funny and unique graphic, and is still in great condition.

COLLECTING AND DEATH

Given these examples, the question remains, why would anyone, myself included, purposefully seek out such Manhattan Project era, Cold War, or post-Cold War items to own or collect? One way of understanding the permanence and longevity of *atomicalia* is to consider the more generic motivations and patterns informing why humans collect, commemorate, and memorialize.

The compulsion to collect artifacts, according to cultural scholars such as Jean Baudrillard, demonstrates an age-old reflection of "the triumph of remembrance over oblivion, to the permanence of Being over Nothingness" where ordering and classifying objects replaces our "desires for suppression and ownership, fears of death and oblivion, hopes of commemoration and eternity."[21] Much of the scholarship in this domain draws from Freudian theories of mourning that observed trauma victims could overcome both the personal loss of loved ones and things, i.e. material objects, by their replacement, restoration, and accumulation.[22] In this context, the collecting of material goods was regarded as a healthy pathological mechanism to compensate for trauma and to help in the mourning process, where objects can be managed by their intercession in daily life (e.g., a photograph, family shrine, keepsake, or memento). Such objectification reminds the subject of their own mortality while simultaneously transcending their fear of death by possessing material that precedes them and will, presumably, "live on" beyond them as inherited or bequeathed goods. For cultural theorist Mieke Bal, collecting "can be attractive as a gesture of endless deferral of death."[23]

So, if these motivations underscore the social conventions of collecting and the general pathology of material possession, what does it mean for collectors/collections embracing the *atomic age*, an era already heavily invested with the symbolic dichotomy of life and mass death? Generically, it is argued, such objects enable their owners to (re)create narratives of a remembered association (a person, thing, place, time, event, etc.) at will by looking at and/or containing that memory around the material item. The permanent and personal display of an object (e.g., on a wall, table, desk, or bookcase), such as an atomic age or Cold War collectible, evokes this association or recollection by its random encounter and intervention in daily life precisely due to its continuous presence. This is unlike an item or specimen that is occasionally returned to through a conscious act of selective choice after its retrieval from a drawer, specimen case, or storage.

It is my contention that collecting has become even more important in the 21st century digital era. Material culture provides a counterpoint to the digital world—its physical, tactile, static, and permanent presence promotes ongoing memorializing, which is something that the fluid, transient, and virtual mass media is incapable of doing. Material culture artifacts have their place, quite literally, within the banality of everyday life, and provide a parallel focus of the routine and mundane, enabling its continued operation through displacement onto an object of remembrance that can then be comfortably "forgotten" temporarily. In this way *atomicalia* might represent a cultural hangover from the *trauma* of "surviving the Cold War," or perhaps it demonstrates the return of the repressed through the negotiation of the contradictions of Engelhardt's "victory culture" via continued material acquisition.

So not all *atomicalia* should be read as fetishistic accumulation and consumption. According to some scholars, kept objects fundamentally serve to stabilize and order the mind.[24] They become increasingly important in times of crisis, upheaval, and chaos, since their permanence, pedigree, and provenance assert survivability across generations. One of the reasons *atomicalia* and Cold War material culture is becoming so attractive may be due to the retrospective recognition of these mass culture objects as totems of the recent past—a past that paradoxically signifies the contemporary occupation of a (previously dubious) future by a post-war generation, prepared for nuclear war—a generation that did not necessarily expect to inhabit any such future. Hence, *atomicalia* reminds us of having lived beyond the unspoken traumas of the Cold War (the ever-present prospect of nuclear oblivion) or the cognitive dissonance of a promised era of atomic plenty (AEC chair Lewis Strauss' "energy too cheap to meter") and a nuclear-powered utopia never achieved.

And yet, *atomicalia* continues to be produced well into the digital, globalized, postmodern 21st century. A stroll through any supermarket or department store will find consumer goods still being promoted with nuclear symbols and icons, or otherwise self-consciously and nostalgically invoking the atomic age in their marketing.

It is deeply paradoxical that the secret $2 billion Manhattan Project and the twin holocausts of Hiroshima and Nagasaki were so quickly exploited to celebrate American victory through popular culture, toys, games, and other forms of entertainment and consumption. As Paul Boyer, Spencer Weart, Allan Winkler, and other American historians have shown, within hours of the Hiroshima bombing, news commentators in the West were expressing the same fears that remain with us today— namely, that the so-called "winning weapon" would be used against victors and foes alike. And we can trace the genesis and evolution of such concerns in the everyday cultural goods we buy and collect.

Notes

1. Michel Foucault, *The Order of Things: An Archaeology of the Human Sciences* (New York: Vintage Press, 1994), xxii.

2. See Susan Pearce, *Objects and Collections: A Cultural Study* (Leicester, UK: Leicester University Press, 1992).

3. See Hillary French, "Linking Consumption, Globalization, and Governance," in the Worldwatch Institute, *The State of the World 2004* (New York: W.W. Norton & Co., 2004), 144–62.

4. Spencer R. Weart, *Nuclear Fear: A History of Images* (Cambridge, UK: Cambridge University Press, 1988).

5. Tom Engelhardt, *The End of Victory Culture: Cold War America and the Disillusioning of a Generation* (New York: Basic Books, 1995).

6. See Raymond Williams, "Culture is Ordinary" in *Resources of Hope* (New York: Verso, 1958; 1989) and *Culture* (London: McMillan, 1977).

7. Norbert Elias, "On the Concept of Everyday Life," in *The Norbert Elias Reader,* ed. Johan Goudsblom (Oxford: Blackwell, 1998), 166–174.

8. Michel de Certeau, *The Practice of Everyday Life*, Steven Rendall, trans. (Berkeley: University of California Press, 1984).

9. Henri Lefebvre, "The Everyday and Everydayness," trans. Christine Levich, Alice Kaplan, and Kristin Ross, *Yale French Studies*, 73, 11.

10. When "atomic" or "nuclear" is included by searching in "item description" the result is over 540,000 hits and 400,000 respectively.

11. Pierre Bourdieu, *Outline of a Theory of Practice*, trans. Richard Nice (Cambridge, UK: Cambridge University Press, 1977).

12. Ben Highmore, *Everyday Life and Cultural Theory* (New York: Routledge, 2002), 17.

13. See Paul Brians, "Farewell to the First Atomic Age," *Nuclear Texts & Contexts*, Fall 1992, No. 8 (https://brians.wsu.edu/bibliographies-discographies-filmographies/nuclear-texts-and contexts-2/nuclear-texts-and-contexts-no-8/); see also Lee Ann Powell, "Culture, Cold War, Conservatism, and the End of The Atomic Age, Richland, Washington, 1943–1989," 2013 (https://research.libraries.wsu.edu/xmlui/handle/2376/5051).

14. See, for example, Paul Boyer, *By the Bomb's Early Light: American Thought and Culture at the Dawn of the Atomic Age* (New York: Pantheon, 1985) and *Fallout: A Historian Reflects on America's Half-Century Encounter with Nuclear Weapons* (Columbus, OH: Ohio State University Press, 1998). See also, Paul Brians, *Nuclear Holocausts: Atomic War in Fiction, 1895–1984* (Kent, OH: Kent State University Press, 1987); Mick Broderick, *Nuclear Movies: A Critical Analysis and Filmography of International Feature Length Films Dealing with Experimentation, Aliens, Terrorism, Holocaust, and Other Disaster Scenarios, 1914–1990* (Jefferson, NC: McFarland & Co., 1991); David Dowling, *Fictions of Nuclear*

Disaster (Iowa City, IA: Iowa University Press, 1987); Joyce Evans, *Celluloid Mushroom Clouds: Hollywood and the Atomic Bomb* (Boulder, CO: Westview Press, 1999); Margot Henriksen, *Dr. Strangelove's America* (Berkeley: University of California Press, 1999); Robert Jacobs, *The Dragon's Tail: Americans Face the Atomic Age* (Amherst, MA: University of Massachusetts Press, 2010); Toni A. Perrine, *Film and the Nuclear Age: Representing Cultural Anxiety* (New York: Garland, 1998); Jerome Shapiro, *Atomic Bomb Cinema* (New York: Routledge, 2002); Allan M. Winkler, *Life Under a Cloud: American Anxiety about the Atom* (Urbana, IL: University of Illinois Press, 1993).

15. *Fallout.*

16. Michael Amundson, "Uranium on the Cranium," in *Atomic Culture*, ed. Michael Amundson and Scott C. Zeman (Boulder, CO: University Press of Colorado, 2003), 49–63.

17. Constadina Titus, "From Political Symbol to Nostalgic Icon: The Kitsch-ification of the Mushroom Cloud," in *Atomic Culture*, 101–123. See also Titus's *Bombs in the Backyard: Atomic Testing and American Culture* (Reno, NV: University of Nevada Press, 2001).

18. See John Fiske, Bob Hodge, and Graham Turner, "Shopping," in *Myths of Oz* (Sydney: Allen & Unwin, 1987), 95–116.

19. *Victory Culture*, 10.

20. Paul Guinnessy, "Components of 'Little Boy' Sold at Auction," 2002. (http://www-personal.umich.edu/~sanders/214/other/news/082002plugsPT.html).

21. Jean Baudrillard, "The System of Collecting," in *The Cultures of Collecting*, ed. John Elsner and Roger Cardinal (Carlton, Vic: Melbourne University Press, 1993), 16, 17.

22. Sigmund Freud, "Mourning and Melancholia," in *The Standard Edition of the Complete Psychological Works of Sigmund Freud, Volume XIV 1914–1916: On the History of the Psycho-Analytic Movement, Papers on Metapsychology and Other Works* (London: Hogarth Press, 1917; 1994), 237–258.

23. Mieke Bal, "Telling Objects: A Narrative Perspective on Collecting," in *The Cultures of Collecting*, ed. John Elsner and Roger Cardinal (Carlton, Vic: Melbourne University Press, 1994), 109.

24. Milhali Czikentsmihaly, "Why We Need Things," in *History from Things: Essays in Material Culture*, ed. Stephen Lubar and W. David Kingery (Washington D.C.: Smithsonian Institution Press, 1993), 20–29.

CHAPTER ELEVEN

Future Directions in Scholarship and Interpretation
A Roundtable Discussion

MICHAEL MAYS [MODERATOR]: We are fortunate to have representatives from the National Park Service here. It's one of the strengths of this conference, that we have people from very different backgrounds speaking to each other; we have different frameworks through which to think about things. Even the formal presentations come out slightly differently. Dan Strom said, "you know this was so much better than a scientific meeting." I wouldn't know much about that, but you do have elements of the scientific meeting, elements of a humanities-oriented academic meeting, people from government, the Park system, and so forth. It's a rich mix.

So, I'd like to preface our discussion by saying I envision this round table as a way to allow everyone, all of us, to reflect on what we've heard over the last couple days in order to identify fruitful pathways or trajectories for future research, future interpretive themes. What are the more striking things we've heard in the various panels and presentations that made you think, "I want to go and do research on that subject" or "I sure would like to see someone else doing research on that topic"?

Bob Clark [WSU Press Editor-in-Chief] and I had a conversation the other day, just a passing conversation about the recruitment of workers at Hanford and the issue of why those workers weren't in Europe or in the Pacific, a point that was raised during one of the panels. What an interesting question. I've had so many of those moments during the conference. I suspect most of us have had several of them. And so I would like to open the discussion this morning by having us identify some of those compelling ideas and questions. We originally framed the conversation, albeit very loosely, as identifying future interpretive themes for the park. But as Kris Kirby [Superintendent of the Manhattan Project National Historical Park (MAPR)] was just saying—and I'll turn this over to her—this is not at all a formal listening session but rather a much more informal discussion and opportunity for an exchange of ideas. And again, what seems to me to have been the strength of this meeting is that we come together—as a community of scholars and professionals with shared interests in the legacies of the Manhattan Project—with MAPR leadership. It's a great opportunity for us, and I hope for them as well, to have important and stimulating conversations about the pressing themes and subjects that haven't thus far received the attention they merit. With that said I'll turn it over to our panel of Park representatives for their opening remarks. And then what I would like to do is open the discussion up to you all, to hear what you have to say and where you think opportunities for future research and interpretation are headed or were you think they should go.

KRIS KIRBY: Okay, thank you Mike for introducing it that way. We, the National Park Service, are really interested in participating in this session. We want to hear your ideas for research, for next steps, et cetera. And when Mike and I talked about how to approach this, we wanted to be very clear that this isn't a formal public workshop for us or a formal listening session. We will be coming back doing that process.

However, since we are here and we have all of these different interests gathered, we're definitely taking advantage of learning from you, from your thoughts. It will help us not only learn more about our jobs and but also help us think about what we want to be doing going forward. I wanted to set the expectation that this in no way is a formal public workshop but that we are going to be listening and just kind of taking it in. Along those lines I wanted to just quickly talk about the rolls of the two agencies [the U.S. Departments of Interior and Energy who co-manage the MAPR]; I'm not going to cite the legislation word for word, but essentially there's a very, very bright line between the roles of our two agencies.

The Department of Energy is responsible for all the physical assets of the Manhattan Project National Historical Park; all the buildings, all the grounds, all the security, the maintenance, all the restoration, everything. The job of the National Park Service is interpretation. We have ultimate authority over all interpretation and we provide an advisory role in historic preservation. I'm just putting this out as a reminder because it's important as we go forward to understand that the pieces that we're going to be taking home aren't going to be, you know; "when are you going to fix the roof on B Reactor?" or "when are you going to get us access to here and there?" Becky and I will be listening for those things that are happening, that seem interesting, that seem we should explore a little more; that is, topics for future conversations when we go back to our offices and start thinking about new directions.

I think the larger piece Mike introduced is that this is really an academic group and it is really exciting to think about, after everything we heard this week, what are the next directions in research and what's the next rabbit trail we need to be chasing in order to figure out where the next level of learning happens? So, I just wanted to emphasize the bright line in legislation and the different roles; I know that's probably the third time you've heard it. But it took me about twenty times to read it to really understand what my job is versus Tracy's versus Colleen's and Becky's.

MM: Thanks Kris. Okay, for those of us in the room who are academics, we now have the unique opportunity to dispense with the pretense of asking questions when what we really want to do is make statements. So, pontificate away. Who would like to begin?

PAUL STANSBURY: I'm Paul Stansbury and I'm currently associated with the Parker Foundation and with WSU, and look forward to helping with the National Park Service. Your bright line in the sand just occurred to me; one of the really good things is that DOE is not responsible for interpretation because if it were, there are people that would criticize it whichever way it went. The National Park Service can say "hey we're independent, we don't have to defend anything." So, I'm extremely excited about the idea that there is a division of responsibilities.

I'm a radiation protection professional and I've already said it, and I'll say it one more time, that radiation protection on an industrial scale was invented here at Hanford. Nobody had ever handled that much radioactive material and basically it was done safely; there are no dead bodies. The Russians, on the other hand, a good fraction of their workers from the early days had radiation pneumonitis—water on the lungs from exposure to plutonium aerosols. Now, we didn't do everything right; we got beryllium wrong while the British got it right, and that sort of thing. Radiation protection is a scientific thing to point out, but as we go forward, the culture that came from the Manhattan Project and surrounding nuclear program is also important. On the more profound side, the field of ecology was essentially invented by the nuclear program. We wouldn't have ecologists today anywhere near as far along if we hadn't developed nuclear technology. Angela [Creager]'s paper pointed to that. And on the more trivial side, I also think of bikini bathing suits as a cultural fall out from—no pun intended—the nuclear program in that this French designer had this hot new bathing suit and then there was the bikini island test, and he just stole the word. So, there are many other nuclear-related subjects and topics that I think would be low hanging fruit for the interpreters to pursue.

WAYNE GLINES: I'll continue the theme. My name is Wayne Glines and I'm also representing the Parker Foundation. I want to echo a comment that Mike and several people have made regarding one of the tremendous benefits of this conference. It has been the ability to look at, understand, and appreciate the many different aspects and perceptions, if you will, of the Manhattan Project and its legacies. Because, coming from a scientific background, I tend to focus on the scientific benefit and outcomes of the Manhattan Project and realizing now there are many other aspects and perceptions, of what the Manhattan Project represented and what its legacies are. In that vein, one of the questions I would encourage for a follow-up conference is, what are the roots of the different perceptions of the Manhattan Project, the Hanford Site, and the nuclear age in general? The psychology, the sociology, the anthropology of the Manhattan Project and the atomic age: To me, those are the areas that link the human side with the technology, and that's something that I would like to see explored, either as part of the interpretive process or certainly in future conferences.

DAVID BOLINGBROKE: Hi. I'm David Bolingbroke, PhD student in history at Washington State University. One thing I would really like to see the park take a look at is, when we think of ecology, especially here at Hanford, we often think of it in a much more recent context. I would love to see some exhibits that talked about, for example, the Lauren Donaldson fish laboratory at Hanford in 1944 and 1945. Or about the work that ecologists were doing as production was starting to ramp up for the first time. Leona Marshall Libby talked about collecting coyotes and then testing their thyroids for radioactivity. So, I think it would be valuable to include ecology in the park's narrative from the very beginning.

DIB GOSWAMI: I have two things to add. One of the things I really liked was the panel on the Manhattan Project in fiction. I mentioned in that session that one of the topics that should be included is that of reality versus fiction. Reality is a little different than what the fiction says. And the second thing is that at many of the Manhattan Project sites we have now gone through several phases of cleanup and environmental restoration. So some coverage of those transitions from weapons production to environmental restoration would be beneficial.

MM: I'll just say that as historians we are always operating in the interstice between perception and reality and that those are perpetually shifting positions. One person's reality is the next person's perception, one generation's hardened realities come to be seen by future generations—sometimes slowly, sometimes quite quickly—as more or less benign perceptions; so I think that's a valuable observation.

John Fox: Yes, well I'm John Fox and I'm representing the B Reactor Museum Association here and our focus, the focus of that organization has been from the start preserving the reactor. But we are in the process of expanding the scope of our endeavor to representing the whole Hanford scope in supporting the Park Service and telling that story. But [Angela Creager's] talk this morning especially struck me in how the radioisotope tracers had an effect on bioecology research. Understanding metabolic processes is something that hadn't occurred to me before, at least. And I just urge the Park Service not to confine the scope of its interpretation of the impact of this project. Because you see how broadly it can reach into areas that we wouldn't think of. And I support the previous comments here particularly in the health physics area, in the whole ecology program which started at Hanford really. It laid the foundation for what we are concerned with today throughout our ecology and what Rachel Carson became concerned about. So, we need to involve a lot of other organizations in support of your interpretation of the story so that you get the whole picture.

Shannon Cram: One thing I think about a lot—I'm a teacher, a professor at UW, and I teach about nuclear studies and Hanford—one of the first things I have students do is; I have them close their eyes and imagine the bomb and draw the first thing that comes to their mind. And without fail, one hundred percent of the time they draw a mushroom cloud and so, I think that the big challenge that you are faced with is to expand the conceptual geography of the bomb beyond the mushroom cloud to all the other complex and disparate impacts and legacies and to not do a before-and-after type narrative. To not be like "the Manhattan Project...and then after the Manhattan Project..." when in fact it is this very fluid continuum that actually never ends. Hanford will never be completely clean; there will always be waste here beyond humanity. To really resist that bounding of time to the Manhattan Project ending in 1946 and to help us understand, as a nation, far beyond the bomb itself, to what the bomb has created. That's what I'm going to be excited to see how you do.

MM: Thank you, and just another quick observation. One of the things that has really struck me over the last couple of days is that we have a great table of young scholars sort of segregated back here in the corner and I don't feel great about that. This isn't specifically focused on the park; I am very happy that the National Park Service has put a focus on the younger generation, fourth graders, kids, and so forth. But I think one of the things those of us who work in academics or in the professions could do a much better job of is mentoring; supporting their research and just generally supporting in any way we can the next generation of scholars and professionals whose work is focused on these sorts of issues. For many of us the Cold War was lived experience. Our younger colleagues are approaching the Cold War from a very different perspective. And personally speaking I couldn't be happier to have you all here; you are highly valued contributors to these proceedings.

CURTIS FOXLEY: Hi, Curt Foxley, Ph.D. candidate from the University of Oklahoma. I don't envy the job here, but I respect it tremendously. I just want to offer some, uh, caution, maybe, because when I think of the [National Park Service] here at Hanford I'm immediately reminded of the *Enola Gay* controversy back in the 1990s and the controversies that are still going on in Pearl Harbor. I can't see into the future, and it's not my job because I'm a historian, but I'm just very concerned about the future as this National Park becomes more popular and gets more press controversy such as the *Enola Gay* and the outrage surrounding that back in the 1990s. It might happen again here, which would be very unfortunate. I'm sure you already know this, but just kind of to advise you to be mindful of that history, of what happened there at the Smithsonian, and try to keep that in mind when going forward, to try to preempt some of this problems that might arise.

KRIS KIRBY: Thank you, and I actually just purchased the book *History Wars* about that controversy for all our locations so that we can be mindful of that controversy. Your point is very well taken, thank you.

TRACY ATKINS: First, just let me respond to what we started with. So, I was with the National Park Service before I joined DOE and worked on the MOA [Memorandum of Agreement] between the Department of Interior and [the Department of] Energy. One of the very first things that we did as the legislation and the agreement were getting signed was host a scholar's forum, and we included the former director of the Smithsonian in that scholar's forum. So, we are well aware that even though it is the Park Service's role now, that telling these stories is complicated and they are not going to be without controversy. But I think that the Park Service folks are the right people to do that.

HEATHER McCLENAHAN: Heather McClenahan, from Los Alamos. The one thing that I'm disappointed about, I'm just sad that nobody from Oak Ridge has been part of this conference because they really are part of the family, and yesterday I called all the Hanford people "my peeps" and I really felt that, so it really has been a great experience. The one thing I want to say publicly, that I've said privately to some of the DOE folks, our friends in Oak Ridge are struggling right now, as you probably all know. AMSI [American Museum of Science and Industry] is being shoved out of its 54,000-square-foot facility down to an 18,000-square-foot bunker. And it's not a happy time and the Department of Energy is going around saying, well, we are not in the museum business, and Eric [Boyle], and Tracy [Atkins], and Padraic [Benson], if you guys would tell your bosses to quit saying that, I would really appreciate it. Because I don't think you could be starting a national park on one hand and saying you are not in the museum business on the other hand. And you know, the history is important to your organization, to our organizations, so if we could all be working towards our history together and not kind of dismissing what is going on with the museums, I think that would be really important.

TRACY ATKINS: I can answer that. So, I would say that the Department of Energy and all the different program offices have an ongoing public education and outreach mission. And that is really where the facility at Oak Ridge is going to be evolving to. The building that AMSI was in had 2.5 million dollars' worth of deferred maintenance and that wasn't really the best way to be spending the dollars versus actually doing outreach. So, that's the driver; there's really good economics for the change there. The Office of Science, the office of the NNSA [National Nuclear Security Administration] folks, they're all working together, including the Park Service, for the plan for the new facility at Oak Ridge, which will be open at the end of December, so, that is really exciting.

JOSH McGUFFIE: Hi, Josh McGuffie, happy graduate student at UCLA. Something I would encourage from the research I've done, and from that of others, is to think about what Hanford is. B Reactor is people's experience of Hanford now, with a lot of sites that are either being built or destroyed. But if you go back and look at maps, and even in our discussions today, it's hard to pinpoint exactly what Hanford is; the 540 square miles, or when did Hanford's boundaries change? So that's part of the story it might be interesting to highlight. But also to highlight what scientists and ecologists at the site thought about Hanford spilling over the boundaries drawn by DOE. You can look at maps of releases from the separation plants and all of a sudden Spokane is part of Hanford because it's in the black part of the map, where there is so much iodine or cesium that has been emitted. So it might be helpful to be thinking critically about the boundaries of Hanford and how they changed, and how they don't always necessarily map on to what DOE or the Manhattan Project said they were.

MM: Interesting idea. And I would just add that there are potentially countless maps that could be generated along those lines. The map of "Hanford" at the Hanford History Project—were we to produce one—would have to be multidimensional and nonrepresentational, because it would encompass not only the physical landscape but also the cultural and imaginative terrains. It wouldn't look anything like the traditional map of the Hanford Site we're accustomed to seeing. The variety of different kinds of maps one can imagine being created is extraordinary. It's possible, or even likely, that with the new technologies increasingly available to us we may be looking at a whole new field of multimodal mapping. Or perhaps that field is already here. In any case, it's a great comment and very suggestive. Thank you.

ANDREW PRINDLE: Thank you. Andrew Prindle from the University of Washington. One of the things I find very interesting in reading the history and hearing this discussion is that Hanford is often offered up as a sort of cornucopia of beginnings. But at the same time, in some of the panels we've heard, there is also the continuation of some fairly uncomfortable things as far as racism, colonial dispossession of native lands, and now irradiated ecologies. So I think preserving the multiplicity of narratives about what Hanford is and what it represents to different people is really important rather than offering up narratives that appear to resolve or contain things in specific ways. The other thing, too, to speak a little bit to Shannon [Cram]'s point, I've increasingly grown to resist the notion of cleanup. You know uranium is not something that you clean up, it's really more a reorganization of waste, and with that said, in my mind at least, is much closer to a suspended risk management program. And I think that making that legible, particularly to younger people because we ourselves and our children and our children's children are going to be inheriting this legacy. Working to make that legible and clear to people, I think, will be very important as younger generations continue to grapple with this environmental and political legacy.

SUSAN SWANBERG: Hi, Susan Swanberg, geneticist and now assistant professor of journalism at the University of Arizona. A number of the commenters have mentioned narratives, and I think it is important to remember that the first draft of history is journalism and not forget William Leonard Laurence and his place. He visited a number of these facilities and wrote about them in his stories and I think there are photographs that exist of him visiting some of these facilities. Photographs of him speaking with Oppenheimer and perhaps some of those things should be part of the exhibits. But I would also like us to consider and think about Laurence and what he wrote and what he did. What we want our journalists to be. Do we want them to be reporters of facts or do we want them to be advocates? And I think that this is a good time to be thinking of those things; what happened with journalism in the Manhattan Project is a good way to segue into the modern issues we're encountering with journalism and facts, with what is true and what is real, and how to report those things. Thank you.

CINDY KELLY: Hi, I'm Cindy Kelly, Atomic Heritage Foundation. I just want to reinforce what Susan said about the importance of multiplicity of views, of looking at Laurence. But also looking at the voices of people who were actually here. I loved Kathleen Flenniken's presentation yesterday because it was her view, but it showed the complexity of any single view of this and the views themselves are so ambiguous and complex. That complexity contradicts the idea that there is a single interpretation available to roll out and be done with it.

What this conference shows is that there are just so many different dimensions and angles. I looked at a paper that a Georgetown law professor did on Gettysburg, and on how that park has been interpreted over the centuries and changed with the prism of the time. There was no talk, as someone mentioned yesterday, until recently, of slavery being the root cause of the civil war. In Wilson's time, the battle was seen as the struggle between these great strong southern boys and those really brave Yankee boys. It was the time of the Progressive Era, the coming of the Boy Scouts, the idea that people should be strong and hardy, the Rough Riders. It might sound bizarre...But it just shows that what we are launching here is something that is going to continue for generations.

I want to compliment Mike [Mays] and Robert [Franklin] and Jillian [Gardner-Andrews] and the others who have put on this wonderful program which rehashes all of the complex issues that are involved. And this is just a group of what, a hundred some people? And to add to that, this is just a fascinating subject that will take our lifetimes and many successive generations' lifetimes to unravel or to deal with. And it has so many lessons for today, as someone was just saying. So, I think [the interpretation of the Park is] going to be terrific, whatever you do is going to be wonderful, and I think that these kinds of programs will complement [that interpretation]. It won't be just the few things you can fit in that eighteen thousand square feet at Oak Ridge. But it will also be other experiences, programs, and interactions that we'll have all across the nation and the world, to try to figure this out. So thank you.

Tom Marceau: Good morning, Tom Marceau, Washington State University. I'm interested from the discussions we've had here in the social dynamics that are taking place during the Manhattan Project. I think that is one of the overlooked dimensions here particularly when we start getting into the roles of men, the roles of women, segregation by class, segregation by race. Not only the types of activities that were put together, all those things that were taking place within the milieu of the 1940s, but how they played out at Hanford with a population pulled from across the country, all thrown into one location here. How that morphed or changed, or what might have happened given those social dynamics of the time. And how all these people came together to actually get this site built. But more than that, what was lurking under the surface, how were they communicating, how were they dealing with each other in those capacities.

Gene Weisskopf: Thank you. I feel the same way about this multiplicity of story lines. It makes the story more complicated, but it makes it deeper and deeper, more deeply interesting every day that goes by. One thing I keep walking away with, one of the ways I've explained B Reactor, is that there is a certain sacredness to it, and I don't mean in the form of one religion or another. I mean in terms of humans understanding the universe in an utterly, utterly, unknown and unpredicted way. And when you look at the reactor you're looking at the face of the universe interpreted by physicists and engineers, to take advantage of something they didn't even know about a few years earlier. What you do with that, I don't know. But I would like people to walk away from the experience with a bit of awe, and a bit of humility instead of hubris and waving the flag, which is another way to do it. So, I don't know how you do that, but in terms of the result of the Manhattan Project I would like to see us offer the city of Nagasaki a permanent display near the reactor, in concert with the National Park Service. But let them tell their story that happened a millisecond after the mushroom cloud and is still happening. The results of the mushroom cloud. I think little children should walk away nervous, knowing that these weapons still exist, and that whoever happens to be in the White House, can decide to use them almost at any given moment. You wouldn't sleep at night knowing that. Thank goodness that can't happen right[?]. Thank you very much, and I don't envy your job, but I am jealous in a little way, and I appreciate what you are trying to do.

ANGELA CREAGER: You've already heard me speak today, so I'll be very brief. I actually wanted to give some first-hand testimony to how significant it was to me, as a historian, to visit Oak Ridge National Lab when I was working on my book. And I was able to go see the old graphite reactor now decommissioned. But it's not really so much open for visit, you have to have an appointment. But it was really terrific for me to see where it was and how it was situated with respect to other facilities at Oak Ridge. To understand that Oak Ridge wasn't one place, but kind of four places and Hanford I know is on an even larger scale. And also the Museum of Science and Industry [AMSI] there [at Oak Ridge], seeing not only the artifacts but the photographs of Ed Westcott. And I think that one of the really lucky things for interpreting the Manhattan Project is that they had such a brilliant and often unacknowledged photographer, and that many of his photographs were at Oak Ridge but not all of them. And the National Archives has many unpublished photographs taken by him and by others and the photographs that I showed from Hanford are all from the National Archives. I think that the challenge for interpretation is that there is rich documentary material available through the DOE open net program and there is rich pictorial material available in the National Archives. But bringing them together as I found in my own work is actually quite challenging. I really encourage you to use the materials not only that DOE has but also the National Archives. And lastly, I'm very grateful that the National Park Service is charged with interpreting this new park facility because I think that the Park Service is a beloved American agency which is not to be taken for granted in any way, anymore. It has had a really important complex and ambivalent history. So, I'm sorry that I wasn't able to visit Hanford when I was writing my chapter that featured some of Hanford's facilities, but I'm really grateful to see it today and anything that I can do, to send you materials, I'd be very happy to do.

DEL BALLARD: Del Ballard here at B Reactor Museum Association and an old Hanford employee. All the discussions and themes that I've heard presented for the Park Service to tell the story of the Manhattan Project have all been in relation to the outcomes, the product, the impacts, the problems, the concerns, the waste, the isotope generation, and all the impacts that those have had. So, it's all been the backside of the outside of the production of the

Manhattan Project. But, what I would like to see addressed is the reason for the Manhattan Project. Why was it created, what was the condition of the world, what were the national security concerns? What were the conditions under which the Manhattan Project was originated, why was it made, why was it put in place, as a theme to be at least addressed? So the public in the future knows the conditions that existed at that time.

TERRY ANDRE: Hi I'm Terry Andre and I am also with the B Reactor Museum Association. But my comment is as a daughter and a mom. I'm a daughter of the WWII generation, the Greatest Generation, I'm a baby boomer and the mother of a millennial, and part of what struck me, through some of the voices in this conference, was that each of these generations and the generations following the millennials has a different experience coming to this story. They have a different cultural background, and I would challenge the academics [to focus in the future on the question of] what lenses do we bring as a WWII generation, as a baby boomer, as a millennial, as the next generation? In particular, I'd be interested in knowing what is being taught in the schools. I do know what is being taught in the local schools—I had to straighten out two local school teachers because the text books were wrong, right here in Richland the text books were wrong—so I'd like to see what is being taught in [terms of] our cultural context, our generational context; it's a really big canvas. I would love to see more academic papers on that in the future.

MM: I think that's a really great comment, Terry, and I want to underscore your point. I think that we academics have some responsibility to not simply or only advance our own research. Don't get me wrong, that research is critical. Again, I look at the table of younger scholars over here, and I look at Angela's work, and I see scholarly work being produced that is awe-inspiring. But we have a multitude of colleagues beyond the walls of the Ivory Tower— just look at the diversity of the panelists at this conference, which so many of you have remarked on—and I think we in higher education sometimes tend to forget our peers and our colleagues in K-12. I think we have an obligation, going forward, to be much more intentional in terms of how we encourage and mentor the next generation of our peers, and to be mindful of the ways we can work together to help students in elementary and secondary schools become more interested and informed about the Manhattan Project and its impacts. And really what that means is curricular development; we don't all have to be in the trenches, necessarily, but we can certainly work with our colleagues in education departments and elsewhere in shaping that curricular development. I'm not talking about the park now; you all have your own essential educational mission. But I think that, to emphasize what Terry was talking about, there are real opportunities to work with teachers and with states to implement curricula that are responsible, forward-looking, and not based on concepts and information that are outdated or, worse, demonstrably wrong.

PAUL STANSBURY: I have another one. One of the sources I think would be really rich for the National Park Service and DOE to explore—and Del [Ballard] from BRMA reminded me of this—is the experience of the worker at Hanford and in particular the bus ride. I'm an avid bridge player and I just realized there's a style of playing bridge where you bid whatever it takes to win on the last hand. And if you go down, you only go down one trick because the game is over and the reason was, the bridge table was something that was carried and put on the bus so that they could play bridge on the way home, and one of the losers had to carry the bridge table home and bring it back the next morning. And so, this is another piece of fallout from tech, cultural stuff that comes from nuclear technology in the weirdest ways.

MM: That's terrific Paul, I love that. And it reminds me of another "Aha!" moment I had last night when Kathleen [Flenniken] was describing her friend Caroline's father with his lunch box getting on the bus to go to work and her own father getting in the car and driving to his office in his suit and tie; it's a very poignant image. And I think, though I'm not an expert in the social history of Hanford, there is not a great deal of research on that milieu. And if that's true, there certainly needs to be much more.

ROBERT WEST: My name is Bob West and I'm from Coeur D'Alene, and I really salute the Park Service. I think what I would like you to bear in mind is that we kind of have a dichotomy: we've talked for three days about all of the wonderful things that went on at Hanford and the new science and everything else. But there's a significant section of the population that want to put the genie back in the bottle, which isn't going to happen. They want to stop making any more nuclear waste because they haven't figured out how to take care of it. And the last thing is that we are proposing not only a cut in the Park Service funds but a $4 trillion upgrade of the nuclear weapons that we have, and that's going to be a very tough thing for the Park Service to handle. I wonder if you could comment about that.

KRIS KIRBY: So we don't typically comment about funding because it's in a different branch of the government, the executive branch of the government, and we don't lobby, we don't work with Congress to give us more money, et cetera. So we don't really comment about funding…obviously we all know more money makes us able to do more things, less money…hinders our ability to do things. But, in general, at the Park Service we've become very creative at leveraging the communities, leveraging academic partnerships, leveraging educational partnerships to help us do our job and kind of spread the story without the level of resources that other agencies might have. So, I don't know if anybody knows this little trivial fact and probably nobody cares except for us. But at every one of the three sites that are in the Manhattan Project National Historical Park [Hanford WA, Oak Ridge TN, Los Alamos NM], the Department of Energy budget at all three of those sites is larger than the entire National Park Service budget.

That being said, the missions are very different and so are we. As someone said earlier, people hold the National Park Service dear, so when we come out to the community and say "Hi, do you want to volunteer? Would you like to help us?" people usually say, "Yes that sounds like a lot of fun." So we do have other avenues, other resources, that aren't necessarily financial-based. We've learned over the years to utilize the funding that is given to us and make the most of it. Thank you for your question.

Afterword

Michael Mays

When the Imperial Japanese Navy Air Service attacked the United States on the morning of December 7, 1941, the research program which would come to be known as the Manhattan Project was in its infancy. Though he had been made aware that German scientists were attempting to develop the world's first atomic bomb as early as August 1939, it was not until October 1941—a mere two months before Pearl Harbor—that President Roosevelt finally approved the United States' own atomic program. In the urgent aftermath of the bombing, as the United States declared war first on Japan and then on Germany, work in the S-1 section of the Office of Scientific Research and Development (the Uranium Committee) intensified considerably. As research into plutonium (first discovered by Glenn Seaborg in February 1941) and nuclear reactor technology began to coalesce at the University of Chicago's Metallurgical Laboratory in early 1942, planning for the massive construction effort was underway in, of all places, a nondescript high-rise office building overlooking City Hall Park in the heart of New York City.

An unlikely location to be certain. But there were several distinct advantages to being there. First, given their experience with management of large-scale construction projects, the Army Corps of Engineers had been chosen to build the production facilities, and the Corps' North Atlantic Division was in the same building. In addition, the offices on the 18th floor, which served as temporary headquarters, were in close proximity to the Manhattan office of the principal project contractor. The location was close to Columbia University, where much of the nuclear research continued to take place. And, finally, there were resources galore, including an abundance of piers to import uranium and other needed ores, warehouses to store them, laboratories to split the atoms, and an unmatched pool of highly-skilled, well-educated military and civilian workers, about 5,000 of whom were employed on the project over time. In June 1942, Colonel James C. Marshall was chosen to head the Army's effort, which was initially named the Laboratory for the Development of Substitute Materials. When Marshall proved ineffectual in his role, he was quickly replaced by Colonel Leslie Groves, who feared the project's original name would draw undue attention to itself. Groves opted instead to utilize the Corps' standard procedure for naming new operations according to the city in which they were located. "Development of Substitute Materials" henceforth became Manhattan Engineer District (MED) and then, in time, Manhattan Project, or more simply yet, "Manhattan."

Unlike other Corps districts, however, the MED had no geographical boundaries. Rightly so, as, despite the many benefits of the New York City location, there were also acute limitations. What Paris had been to the eighteenth century and London to the nineteenth, so Manhattan was to the first half of the twentieth century: the dynamic global center of finance and fashion, art and architecture, business and industry; the bustling, vibrant place to be. "The place to be," however, was singularly unsuitable for a top-secret government project requiring vast amounts of space, abundant natural resources, and substantial population-free buffer zones to mitigate safety and security concerns. Hence, Groves aggressively went to work identifying and securing the sites to which "Manhattan" would be dispersed: Oak Ridge, Tennessee, where uranium was processed, in September 1942; Los Alamos, New Mexico, where much of the research was conducted and the bombs assembled, in November 1942; and Hanford, Washington, where plutonium was produced, in February 1943. As these remote and far-flung nuclear enclaves grew, so Manhattan's role shrank. In 1943, just a little more than a year after it was established, the Manhattan Engineer District headquarters were relocated to Oak Ridge. Yet even after this geographical dislocation, Manhattan, through its scientists, its businessmen, and its journalists, retained a role in the unfolding atomic drama. The story of Groves hiring Oppenheimer to direct Project Y is well known. Less familiar is another hire Groves made, that of *New York Times* science writer William L. "Atomic Bill" Laurence, whose controversial career is the subject of Susan Swanberg's "Borrowed Chronicles." The Pulitzer Prize-winning Laurence, as Swanberg recounts, was hand-picked by Groves to document the Manhattan Project and would become the only journalist to witness both the Trinity test and the bombing of Nagasaki.

Nevertheless, while some individual New Yorkers continued to play minor parts in the war-time effort, the city itself had clearly ceased to be the focal center of the Manhattan Project. The decision to physically relocate the nuclear program had been purely pragmatic—straightforward and necessary. The ontological decentering that the project would eventually engender, however, was of a different, far more complex order altogether: shattering core beliefs and values. That crisis would come soon, and without advance warning. For the moment, as Laurence departed New York for Los Alamos in April 1945, Manhattan remained the brash and confident epicenter of the civilized world.

It is unlikely many of the writer's fellow cosmopolites could easily have located New Mexico on a map. But there, in the desert near Alamogordo, "in a burst of flame that illuminated earth and sky for a brief span that seemed eternal," Laurence witnessed the birth of the Atomic Age.[1] Though few were aware at the time, nothing would ever be the same again. By the time he returned to New York just a few short months later, the whole world would know the news, largely thanks to Laurence himself, who reported on it for the *Times*. The consequence of that news—what it meant—was much more difficult to convey or digest in the immediate aftermath. But in Laurence's heraldic journey from Times Square—the very center of the center—to the New Mexico periphery, we glimpse for the first time the epistemic shift wrought by the new nuclear age: the transformation from a spatially structured world measured by size and scope and scale and strength, to a temporally structured world measured by the sophistication of one's technologies and characterized by the always-imminent threat of apocalypse. In this reordered universe, the concept of the geographical center—the global capital—became obsolete, another of Hiroshima and Nagasaki's many casualties. As Laurence had been one of the first to understand, suddenly, in that eternal instant of Trinity's burst of flame, Manhattan had become the past while "Manhattan" presaged the future.

In hindsight it may appear self-evident to contend—as Laurence did (or seemed to), and as I do here—that the Manhattan Project was the most significant event of the twentieth century. Yet before moving on we might recall, briefly, some other important occurrences that profoundly shaped that age (listed here in no particular order): The precipitous decline of centuries-old empires and the sudden proliferation of countless new nation-states. Two world wars fought within a generation of each other. A nearly half-century-long cold war which fostered plenty of hot ones fought in proxy. A worldwide economic depression. A global civil- and gender-rights movement. The creation of the assembly line. The mass production of the automobile. The industrialized murder of the Holocaust. The development of air and space travel. The inventions of the personal computer, the internet, and the cell phone. And, not least, the death of God.

To designate as especially important just one event from that list strikes me as anything but an obvious choice. On the contrary, in the context of these prodigious changes, hallmarks of the last century's vast global upheaval, to single out the Manhattan Project for its significance is a bold claim indeed. But it is one the essays collected here—in their breadth and depth and scope—convincingly and repeatedly bear out.

"Among the calamities of war," Samuel Johnson wrote in one of the many *Idler* columns he penned between 1758 and 1760, "may be jointly numbered the diminution of the love of truth, by the falsehoods which interest dictates and credulity encourages." Johnson's scorn was directed at a new breed of writer, the journalist, whose primary purpose as he saw it was to provide "amusement for the rich and idle." The two defining characteristics of these "newswriters," as he derisively called them, were "contempt of shame and indifference to truth." And wartime, Johnson went on to claim, offered the perfect opportunity for the application of these traits. It is unclear whether Hiram Johnson, a progressive Republican representing California in the U.S. Senate during the First World War, had ever read the brilliant polymath with whom he shares a surname. But the observation first attributed to Senator Johnson in 1918—that "the first casualty when war comes is truth"—certainly echoes Dr. Johnson's view, albeit far more prosaically. While journalists and the practice of journalism have changed profoundly since the eighteenth century when the profession was in its infancy, the news business continues to serve as a sort of Rorschach test in which we discover what we wish to see.

Now, seemingly more than ever in this age of social media with its real and perceived biases, and as we are inundated with claims and allegations of "fake news" and Russian hacking, it is all too easy to find oneself bewildered and perplexed in the face of the instantaneous and relentless 24/7/365 barrage of so-called "news."

Yet as Hillary Dickerson and Susan Swanberg remind us in Section I, which takes its title from the latter-day Johnson's pithy quip, journalism of every stripe has always been a magnet for controversy, an object (to paraphrase one of its own) of "fear and loathing."[2] It is commonplace in times of war or other severe political crises that governments, anxious about security or control of their messaging, resort to propaganda and, in more extreme cases, to outright censorship. In this respect, Dickerson's chapter, "Atomic Legacies in Censored Print: Newspapers and the Meaning of Nuclear War," tells a familiar story of military control of the press to shore up the party line. The fact that Japanese militarist censorship gave way to Allied Forces' Occupation censorship in the weeks and months following the war is unremarkable. That the inverted narratives of that propagandizing so closely resembled each other is equally unsurprising. What Dickerson discovers in her careful analysis of two English language Japanese dailies, the *Mainichi* and the *Nippon Times*, is both a "briefly free space" in the transition between the old and new state censorship and, more notably, traces of a counter-narrative under Occupation that, despite the latter's stringent control, still managed to question American beliefs in the just nature of atomic destruction. "Japan's English-language dailies reflected the United States' fascination with the power of its atomic creation," Dickerson concludes. But they did so in a way that suggests a degree of "Japanese agency in crafting atomic narratives" that ultimately served to "subvert American claims of righteous victory" and "complicate national narratives of the Asia-Pacific War."

The story of William L. "Atomic Bill" Laurence is also one of propaganda. Laurence, as noted earlier, was reporting on science for the *New York Times* when Groves chose him to become the official historian of the Manhattan Project. Laurence earned the second of his two Pulitzer Prizes for his 1945 coverage of the atomic bomb beginning with the eyewitness account from Nagasaki. Controversial even in his own day, the controversy surrounding Laurence and his reporting intensified following the publication of *Hiroshima in America: Fifty Years of Denial* in 1995. The book's authors, Robert Jay Lifton and Greg Mitchell, denounced Laurence as a government lackey: "Here was the nation's leading science reporter, severely compromised, not only unable but disinclined to reveal all he knew about the potential hazards of the most important scientific discovery of his time."[3] In the wake of these revelations, the journalists Amy Goodman and David Goodman went even further, calling for the Pulitzer Board to strip Laurence and the *Times* of the 1946 prize. "In his faithful parroting of the government line," they concluded, Laurence "was crucial in launching a half-century of silence about the lingering effects of the bomb."[4]

In its breach of the most fundamental of journalistic ethics, Laurence's behavior raises a host of questions regarding journalists' relationships to their subject matter, particularly in times of war. But Laurence's already ethically-muddied story gets considerably more complicated as a result of Laurence's tendency to "borrow"—often extensively—from others' writing (and his own) without proper attribution. In "Borrowed Chronicles: William L. 'Atomic Bill' Laurence and the Reports of a Hiroshima Survivor," Swanberg carefully traces Laurence's previously unrecognized patterns of plagiarism, placing it in its historical, legal, and ethical contexts. In his two books about the atomic bomb, *Dawn Over Zero* and *Men and Atoms*, Laurence appropriated (without proper citation) substantial passages from his own previously published *Times* stories and from the first-hand account of the Hiroshima bombing written by Father John A. Siemes, S.J. As every teacher knows, plagiarism is a sticky wicket and Swanberg is right to ask "whether criticizing [Laurence's] imperfect acknowledgment of Siemes' work is presentist[?]" The answer, she suggests, is complicated. Laurence had no formal training in journalism, the overlap between copyright law and plagiarism at that time constituted a legal grey area, and journalistic ethics of the period were evolving. And yet.... Careful scrutiny of Laurence's transformation of Siemes' first-person account into a third-person chronicle indicates, she concludes, "a level of engagement with the material that belies mere carelessness," and thus that his textual manipulations are intended to make his readers "think he, Laurence, wrote portions of Siemes' moving account of the devastation endured by the city of Hiroshima and its residents."

The essays in Section I underscore the need to maintain a vigilant and healthy skepticism regarding what we read in print or consume through multiple and various forms of media. They also provide a bracing reminder—so timely in the present moment—of the inestimable value of a free and open press. And thus they provide a broadly cautionary tale about the risks posed to a democratic society when its institutions, and the people who operate within them, start down the slippery slope of deviating from the truth and begin to make things up, so to speak. Propaganda, censorship, plagiarism: each is a little executioner of truth, and never more so than in times of great crises.

The essays in Section II, "Necessity is the Mother of Invention," are also preoccupied with the concept of "making things up," albeit in a very different way. Each of the essays here focus on the distinctive ways in which crises—in this case the requirements of the Manhattan Project and the emergence of new social and political formations in the United States following and deriving from the war—generate constructive or creative responses. Crises are, by definition, turning points or watershed moments when an old order is being upended by new and unprecedented circumstances, conditions, realities. They demand innovation, adaptation, resourcefulness, industry, creative problem-solving, and thinking outside the box. They are the unstable foundations that generate, of necessity, new orders, new ways of knowing, new social formations. They are likely to be experienced contemporaneously as chaotic, turbulent, unsettling, or even calamitous. Yet once a crisis is sufficiently in the past, and with the benefit of hindsight, we come to recognize the fecund energies they produce. The Manhattan Project was the United States government's response to just such a crisis: the crisis of attempting to harness and utilize a largely theoretical new technology before Nazi Germany could. And its effects were revolutionary. As Ellen McGehee states simply, "The wartime program resulted in science and technology that transformed the role of the United States in the world and ushered in the atomic age." The four essays in Section II illustrate the pivotal role the Manhattan Project's scientific and technological innovations—and, just as importantly, the conditions of their creation—played in the sweeping restructuring of social and political formations in the new post-war global landscape.

McGehee's "Casting Shadows, Capturing Images: The History and Legacy of Implosion Physics at Los Alamos," begins—like a classic Greek tragedy—*in medias res*, in the midst of Los Alamos's frantic race to develop an effective implosion method for the plutonium-based Fat Man atomic bomb. Those working at the secret Project Y laboratory confronted one of the Manhattan Project's greatest technical challenges: "the unsuitability of the plutonium gun device and the need to develop an alternate weapon design." As with most every other aspect of the Manhattan Project, the Los Alamos scientists and engineers based their work on what little they were certain of and made up the rest on the fly, through a process of trial and error. The so-called "Crisis of '44" exemplifies the type of difficulties the architects of the program faced. As McGehee reports, in 1944 the Los Alamos site suffered a devastating setback when scientists discovered that plutonium could not be used in the Thin Man prototype they had been developing and would have to be abandoned. In response, "J. Robert Oppenheimer reorganized Project Y ...with one goal in mind: to develop an implosion or Fat Man-type weapon to make use of Hanford's plutonium." The need to develop new technology suited to Hanford's plutonium rather than Oak Ridge's uranium resulted, McGehee notes, in a whole new field of scientific inquiry, that of implosion physics.

New fields and new formations are the theme of Section II and in "Herbert M. Parker, Health Physics, and Hanford," Ronald Kathren provides an account of the career of one of the leading pioneers of not one but *two* new fields: medical (or radiological) physics and health physics. Before the war, health physics did not exist and radiation safety was of only minor concern, Kathren observes; however, "the establishment of the Manhattan Project to build an atomic bomb changed all that." Plutonium itself was a very recent discovery and—apart from the fact that it had characteristics similar to radium—little else was known about this man-made element or the risks it posed to human health or environmental safety. Fermi and Groves both expressed concerns, the latter emphasizing the need to take every precaution to safeguard the health of Hanford workers. On the recommendation of Nobel Laureate Arthur Holly Compton, director of the University of Chicago's Metallurgical Laboratory, the English-born Parker was recruited from Oak Ridge to lead the radiation safety program at Hanford. Once there, he made numerous seminal contributions to the practice of health physics, including devising a system of dose quantities and units that is the basis of our modern system: "Many of the operational techniques used to limit exposure and ensure worker protection against radiation exposure," Kathren writes, "were developed and remain to this day a part of the armamentarium of professional health physicists." Nor did Parker's influence wane after the Manhattan Project; later in his career he was an outspoken advocate for the need to develop radioactive waste management efforts, yet another new profession to which he contributed many of the basic underpinnings.

The new fields of implosion physics and medical and health physics were necessary responses to unique challenges faced by Manhattan Project scientists and engineers as they raced to fulfill their mission of constructing a nuclear weapon. But after the success of the Trinity test and the atomic bombings of Hiroshima and Nagasaki, and with a Cold War turning increasingly hot, the proverbial genie of nuclear fission was out of its bottle and there would be no putting it back. Hence, following the end of the war another entirely modern—and heretofore unimagined—industry was born. How, and who, would manage the new nuclear technologies? As David Munns argues in "'The Atom Goes to College': The Teaching Reactors that Trained the Atomic Age," much of that work would be shouldered in the post-war years by the nation's colleges and universities. During the height of the Cold War, from 1949 to the late 1970s, nearly 150 university programs and more than two dozen reactors sprung up on campuses across the country, Munns notes, adding that their emergence "represents one of the key legacies of the Manhattan Project." Those programs taught the advanced technical skills and knowledge needed to supply the new experts of the post-war era, to be sure. "Undergirding the multiple infrastructures of weapons, power plants, and radioisotopes were legions of trained nuclear engineers, scientists, technologists, and technicians." But the programs did much more than simply train a vast new labor pool, Munns contends: they instilled the values and habits of mind of the new Atomic age. In learning "how to operate the complex economic, social, and military infrastructure of the state that always threatened to run out of control," what students gained from their education and experience above all, he writes perceptively, "was the knowledge and assurance that they could administer their modernist state."

From an institutional standpoint, universities have always been at least as much concerned with inculcating socially normative values as they have been with teaching critical thinking or the scientific method. So it should come as no surprise that, with a post-war technocratic American hegemony in the ascent, its system of higher education would quickly have adapted to and come to reflect the needs and values of the new age. As crucibles of shifting beliefs and practices, universities mediate the dynamic interplay of individual and collective values through which any given "community" is formed. They are not alone, of course; family, churches, political parties, trade unions, and countless other institutions, organizations, and associations all serve a similar role.

The forging of one these unique communities—a politically engaged and activist atomic scientists' movement—is the subject of Ian Graig's "Political Scientists: The Atomic Scientists and the Emergence of a Politically Engaged Scientific Community." Tracing the genealogy of the Earth Day 2017 March for Science—when thousands of scientists in Washington and other cities around the country called on the Trump administration to support the sciences and take action on pressing environmental and social issues such as climate control and vaccines—Graig finds its roots in the political activism of the Manhattan Project's atomic scientists. Before World War II, he argues, "the scientific community in the United States had little history of political engagement." Ironically, it was the atomic scientists' very success that catalyzed their political activism, having witnessed first-hand the destructive powers of the technology they had created. Most of the politically active scientists shared the same unwavering belief in the power of science to do good. But "neither faith in science nor belief in the importance of defeating Germany and Japan could overcome the Manhattan Project scientists' fears that development of atomic weapons posed a very real threat to humankind," especially when left in the hands of the generals and politicians. Where previous generations of scientists had rarely ventured into the political arena— and even then, only tepidly—the Manhattan Project scientists "jumped whole-heartedly into the world of politics," Graig writes. "They lobbied members of congress directly, testified at congressional hearings, wrote reports and opinion pieces, published newsletters, and organized conferences and rallies." Moreover, they "showed a willingness to engage in grassroots politicking in a manner more reminiscent of a ward boss than a group of physicists and chemists." In engaging the public and policy makers on a broad range of issues that went well beyond the sciences, the atomic scientists broke with the tradition of political quietism of earlier generations of scientists and paved the way for the continued political engagement of the American scientific community. The atomic scientists' political activism marks a critical turning point, Graig concludes, and thus remains a vital and enduring legacy of the Manhattan Project.

The essays in Section II each describe a particular problem or challenge requiring the creation of something new (new fields of knowledge, technologies, skills, training, education, outlook—whether all together or in some combination). The solutions to each of those unique problems led in turn to important advances in existing fields (education, science) or to the creation of entirely new fields (implosion diagnostics, health physics). The legacies outlined here, then, appear to confirm the prevailing spirit of the pre-Cold War era, as expressed by Niels Bohr (and quoted in Graig's "Political Scientists"). Writing just days after Hiroshima, Bohr argued that the discovery of nuclear fission must inevitably prove to be a force for good since "the progress of science and the advance of civilization have remained most intimately interwoven." Yet these remedies, remarkable as they were, also register ongoing deeper-level disturbances. Parker's discoveries in health physics were so important because radiation is a peril to both humans and the environment. The atomic scientists became politicized because they were concerned about control of their new technology. And a burgeoning nuclear stockpile was a Cold War necessity, but how would it be managed? McGehee, too, focuses on the impediments Laboratory Y scientists and engineers had to surmount, and the remarkable ingenuity and industry they displayed, in successfully developing the new implosion technology needed for a plutonium-based weapon. But "Casting Shadows" does not end with Trinity, or Nagasaki, or the Cold War even. Instead, it follows its story through time, up to the present day, uncovering in the process many of the neglected or overlooked costs associated with that remarkable achievement. Little remains of the wartime laboratory at Los Alamos. But as a trained archaeologist McGehee is skilled in the art of reading forsaken landscapes which "retain historical significance" and can "serve to stimulate meaningful dialogues about the past, its memory, and meaning." The historical vestiges the archaeologist disinters from Los Alamos's scientific landscapes, McGehee asserts, are indispensable puzzle pieces as the new national park sets out on its mission of attempting to tell the whole story of the Manhattan Project.

In guiding us back to a number of the park's own more vexing interpretive themes, including the displacement of local populations, worker experiences, and the human and environmental consequences of the wartime effort, "Casting Shadows" serves to remind that no monumental achievement is ever accomplished without costs, often substantial ones. Some of these—the Manhattan Project's darker legacies—are front and center in Section III, "Facts and Fictions." Here, Daisy Henwood and Laura Harkewicz ruminate upon the physical and psychological toll of post-war nuclear testing while Michelle Gerber documents the mostly unsuccessful efforts of Hanford scientists, during its production years, to limit the chemical contaminants entering the Columbia River and its ecosystems. The extent to which "uncertainty" asserts itself as a prevailing psychic mode in the literal and metaphorical post-war landscape is a consistent theme throughout the section. The uncertain correlation between exposure and disease, for instance, or the confusion inherent in the slippages of translation, were nearly as distressing to the afflicted as were the afflictions themselves. That dubious science was often invoked by government officials to obfuscate unpleasant or inconvenient realities, heightened that uncertainty and thereby intensified the frustration and anger victimized communities experienced. To "Facts and Fictions," then, we might add a subtitle modified from Section I: Trust is the first casualty of government secrecy.

For all intents and purposes, the bombings of Hiroshima and Nagasaki brought the war to a close. Yet—reiterating a dominant motif of this collection—the significance of those two events rests far less in the ending they signaled than in the beginning they portend. As discussed above, in the New Mexico desert near Alamogordo the Trinity test secretly gave birth to the new atomic age while the public announcement followed soon after, in August 1945. Less than a year later, with Operation Crossroads, the United States tested a pair of nuclear weapons at Bikini Atoll in the Marshall Islands. Initially the pace of testing was relatively slow, largely because its arsenal was limited. But when the Soviet Union unexpectedly conducted a successful nuclear test of its own in August 1949, the arms race that would consume the next four decades was officially on. From the beginning of the testing program in 1945 through its cessation in 1992, the United States conducted 1,045 nuclear tests. The majority of these took place first at the Pacific Proving Grounds and other locations in the Pacific (106 tests in all) and later at the Nevada Test Site some 65 miles northwest of Las Vegas (904 tests). The Brookings Institution, in its "U.S. Nuclear Weapons Cost Study Project," estimates the minimum incurred costs of the U.S. nuclear program from the start of the Manhattan Project through 1996 to be $9.3 trillion (in 2018 dollars).[5] During that same period it is estimated to have produced more than 70,000 nuclear warheads, a figure greater than all other nuclear states combined. The full impact of the nuclear program on human health and the environment has been notoriously difficult to ascertain, and will most likely never be known, as each of the essays in Section III is at pains to stress. What we do know for certain is that, as of 2018, more than $2.25 billion in damages had been awarded to claimants through the Department of Justice's Radiation Exposure Compensation System. Yet, as Daisy Henwood argues in "Pursuing the Cancellation of the Apocalypse: Terry Tempest Williams' *Refuge* and Rebecca Solnit's *Savage Dreams*," these figures, staggering as they are, tell only a small part of the story of the aggregate toll of the new nuclear age and its Cold War arms race.

Engaging the work of two contemporary U.S. women writers, Henwood eschews the typical subject matter of so much Manhattan Project-related research—the American postwar political landscape, Cold War intrigues, nuclear proliferation, environmental cleanup, and so on—choosing instead to focus on the often ignored social and local impacts of weapons development. Looking specifically at nuclear testing in the Nevada desert, Henwood finds at the center of both texts an interpretive impasse, or "ambivalence," she sees as an intrinsic characteristic of the nuclear age. This condition of "not-knowing" is best illustrated in Williams' *Refuge*, where the writer's family has been so ravaged by cancer that she finds herself its matriarch at the age of 34. Of Williams' reluctance to draw a causal link between nuclear testing and her mother's illness, Henwood writes: "the ambivalence…marks an epistemological paralysis in which Williams both knows and does not know, is both sure of and yet persuaded against the connection between her mother's illness and nuclear testing." This "toxic anxiety," she contends, is both a product of and exacerbated by governmental dissembling—aided and abetted by an evasive (pseudo-)scientific discourse—surrounding the risks posed by the nuclear testing program. For Henwood, this ambivalence, or indeterminacy, is the new condition of reality: vacillating between acceptance and denial of the nuclear world's ubiquitous dangers we exist in a paralyzing state of uncertainty. Paralysis is the bleak underpinning of both texts. Yet for both writers, growing awareness elicits a third, more hopeful, response outside the debilitating either/or impasse—"activism." Williams' and Solnit's respective journeys lead to their commitment as activists. As Henwood concludes, however, the force of that activism resides at least as much in their words as it does in their deeds. For in their determination to move beyond "not knowing," and hence to begin to think the unthinkable, their writings "not only describe action, they provoke it, by calling up our feelings, our outrage. Writing is their way of pursuing a future beyond the apocalyptic promises of nuclear technology."

Activism is a central concern of Laura Harkewicz as well, though the activism she describes is decidedly one of acts, since words, as she argues in the case of the Marshallese Islanders, served to impede rather than provoke action. In "'We Can't Relocate the World': Activists, Doctors, and a Radiation-Exposed Identity," Harkewicz reports on the experiences of the Rongelap and Utirik peoples who were inadvertently exposed to radioactive fallout in 1954 when the yield of the United States' first hydrogen bomb, Operation Bravo, proved to be two and a half times stronger than predicted. As with Graig's politicized atomic scientists, who actively forged a new type of scientific community in response to shifting political circumstances, so the Marshallese "mobilized around a radiation-exposed identity to claim membership in a community of individuals who had suffered similar radiation-induced injuries." While the Bravo Medical Program was put in place to treat the victims of the fallout, and continued for 40 years, the Marshallese—or the antinuclear activists acting on their behalf—continued to question and cast doubt upon the motivations and interests of the program and its doctors. As Harkewicz points out, Bravo administrators made virtually no effort to assuage these concerns. Doctors "failed to learn the language of the Marshallese," "never translated their reports," "never took the time to explain," and "dismissed complaints they did not expect as psychological." In aligning themselves with other radiation-exposed peoples—atomic veterans, U.S. servicemen, downwinders—the Marshallese sought not only to justify their claims for financial compensation and medical care but also "to challenge national laboratory scientists' 'monopolistic claim over knowledge'." The story Harkewicz narrates, as she herself readily acknowledges, is shrouded in uncertainty. Activists used the Marshallese's plight to advance their own antinuclear agenda while Bravo doctors and scientists carried out their work with no apparent accountability to anyone but themselves. Her story, then, situates itself squarely in the same space of indeterminacy or ambivalence Henwood describes so well.

Whether motivated by her years spent witnessing uncertainty first-hand or, perhaps, opting for her own third way, Michele Gerber, in "Hanford Production Reactor Operations and Contamination in the Columbia River," presents a bracingly fact-based report on the inability of scientists to mitigate the release of radiological and chemical contaminants into the Columbia River, which severely affected aquatic life and the entire food chain. To cite just a couple of Gerber's examples indicating the scope of that production and the challenges it wrought: Between 1944 and 1963 nine nuclear reactors were brought online at Hanford. During that time water usage by all reactors per day increased from 1.29 billion gallons in 1945, to 5 billion gallons in 1955, to 9.76 billion gallons *per day* in 1964 when production peaked. Increased production meant more and more contaminants in the effluent discharges. But it also meant increasingly higher temperatures of the water being returned to the Columbia and its ecosystem: from 149°F. during World War II, to 194°F. in 1956, to 203°F. in 1964. "Increased river temperatures," one report stated, "constitute the greatest potential threat to the local salmon." Despite their repeated efforts, and the implementation of a broad range of strategies, there was little the Hanford scientists could do, Gerber argues, but wait for reactor operations to abate. And remarkably, "[b]y the late 1970s, the volume of effluent discharges to the Columbia River stood at less than two-tenths of one percent of 1964 high levels."

It would be satisfying to end this afterword on that bright note. But appearing to imply everything is better now would also be duplicitous.

It would require ignoring current global political tensions surrounding the question of which countries will or will not be allowed to be nuclear states. Closer to home, it would require ignoring the ongoing $2 billion-plus per annum cleanup at the Hanford Site, a mission many expect to continue for at least another half-century. Most immediately, it would require overlooking the complex legacies—for good and ill—the essays gathered here so assiduously describe. As they attest, we live still in the immense shadow of the Manhattan Project. If that past still shapes us, however, the futures we imagine and create will be of our own making. In accordance with Kierkegaard's observation that "life can only be understood backwards; but ... it must be lived forwards,"[6] it seems fitting that the collection concludes with a similarly Janus-faced orientation, Section IV's "Looking Back, Looking Forward."

There, in "*Atomicalia*: Collecting and Exhibiting Manhattan Project Material Culture," Mick Broderick reflects upon his own experiences as a collector of Manhattan Project memorabilia. His collection formed the foundation of an exhibition he curated for the state museum of Western Australia titled *Half-Lives: Experiencing the Atomic Age* (and subsequently repurposed as *Atomicalia* for exhibitions in Japan and Canada). As Broderick demonstrates, nuclear iconography is ubiquitous. An abbreviated list of products utilizing atomic imagery includes razors, deodorant, salt and pepper shakers, popcorn poppers, energy juices, lighters, cuff links, belt buckles, boxer shorts, and on and on. Nor are these objects solely the products of Cold War-era kitsch; as Broderick notes, "they are as abundant and prolific today as they were in the 1950s through the 1980s." The question "*Atomicalia*" poses is why nuclear objects continue to be produced and enjoy the popularity they do even now, nearly two decades into the twenty-first century. There is no single or simple answer to this question, and Broderick does not presume to offer one. But contemplating that appeal he offers a number of insightful observations both on the nature of collecting in general and on our continuing preoccupation with nuclear iconography in particular. Kept objects are especially important during unsettled times, Broderick suggests, "since their permanence, pedigree, and provenance assert survivability across generations." For a postwar generation who came of age anticipating an always-imminent apocalypse, the present is a gift, a future that generation did not expect to happen. Thus *atomicalia*, Broderick speculates, "reminds us of having lived beyond the unspoken trauma of the cold war."

The fascination with *atomicalia* might indeed be the product of a certain talismanic aura. But in the same sentence Broderick considers an alternate and seemingly contradictory possibility. *Atomicalia* may invoke for us instead "the cognitive dissonance of a promised era of plenty...and a nuclear-powered utopia never achieved." Does our ongoing obsession with the atom represent tragedy averted, then, or promise unfulfilled? Both? Neither? That uncertainty, that ambiguity, is yet another of the underlying themes of this collection. What is the legacy of the Manhattan Project three-quarters of a century later? As both these essays and the roundtable discussion concluding the volume make plain, no single legacy is definitive. That story is, well...complicated. The challenge—and the opportunity—for the National Park Service with its new Manhattan Project National Historical Park will be to draw together these complex, controversial, and often conflicting legacies as it begins to piece together the comprehensive and inclusive story of the most important event of the last century.

Notes

1. "Eyewitness Account of Bomb Test," *New York Times*, September 26, 1945.

2. One need only think, for example, of Richard Nixon's self-pitying admonishment of reporters in the wake of his stinging loss in the 1962 California gubernatorial election: "You don't have Nixon to kick around anymore."

3. Robert Jay Lifton and Greg Mitchell, *Hiroshima in America: Fifty Years of Denial* (New York: Putnam, 1995). Quoted in Amy Goodman and David Goodman, *The Exception to the Rulers: Exposing Oily Politicians, War Profiteers, and the Media That Love Them* (New York: Hyperion Press, 2004), 296–298.

4. Amy Goodman and David Goodman, "Hiroshima Cover-up: How the War Department's Timesman Won a Pulitzer," *Common Dreams*, August 10, 2004.

5. http://www.brookings.edu/about/projects/archive/nucweapons/figure1.

6. Søren Kierkegaard, *Papers and Journals*, ed. and trans. Alastair Hannay (New York: Penguin, 1996), 63.

Contributors

Mick Broderick is an associate professor of media analysis at Murdoch University in Perth, Australia. A specialist in nuclear culture and its media representation, he is author of over 100 scholarly outputs including research monographs, journal articles, book chapters, curated exhibitions and digital media productions.

Hilary Dickerson is a professor of history at Walla Walla University, where she teaches classes on the United States and Modern East Asia. Her research interests include the atomic bombings of Hiroshima and Nagasaki, the American occupation of Japan, the Asia-Pacific War, and transnationalism.

Michele S. Gerber has consulted nationally and internationally to government and private entities, including the U.S. Centers for Disease Control, the General Accounting Office, Washington state, and the government of Japan. She wrote the technical volume that resulted in B Reactor becoming a National Historic Landmark and authored a best-selling book on the Hanford Nuclear Site that has been published in four editions. She retired from the Hanford Site in 2011, and is now the president of the Benton Franklin Recovery Coalition.

Ian Graig is the chief executive of Global Policy Group, a public policy consulting firm based in Washington, DC, which he helped establish in 1995. He previously held positions with SRI International and the Washington law firm of Tanaka Ritger & Middleton. He has written extensively on U.S. energy, environmental, climate, technology, and foreign policy.

Laura J. Harkewicz is a lecturer in the Interdisciplinary Arts and Sciences Program at the University of Washington Bothell.

Daisy Henwood is a third year Ph.D. candidate at the University of East Anglia in Norwich, England. She looks at ecofeminism in Rebecca Solnit's work, with a focus on writing as climate activism.

Ronald L. Kathren is professor emeritus in the department of pharmaceutical sciences at Washington State University Tri-Cities, where he also served as director of the United States Transuranium and Uranium

Registries. Along with numerous journal articles and books, Kathren's influential publications include *Herbert M. Parker: Publications and Other Contributions to Radiological and Health Physics* (1986) and *The Plutonium Story: The Journals of Professor Glenn T. Seaborg 1939-1946* (1994).

Michael Mays is professor of English and founding director of the Hanford History Project at Washington State University Tri-Cities. He is the author of *Nation States: The Cultures of Irish Nationalism* (2007) and co-author of *World War I and the Cultures of Modernity* (2000). He serves as editor for Washington State University Press's Hanford Histories book series.

Ellen D. McGehee, Ph.D., is a historian at Los Alamos National Laboratory. Her primary fields of study include New Mexico homesteading, the women of wartime Los Alamos, and the history and built environment of the Manhattan Project and Cold War eras.

David Peter Dell Munns is associate professor of history at the John Jay College of Criminal Justice, CUNY, in New York City. Specializing in the history of Cold War science and technology, Munns is the author of *A Single Sky: How An International Community Forged the Science of Radio Astronomy* (2013), *Engineering the Environment: Phytotrons and the Quest to Control Climate in the Cold War* (2017), and is a co-author of *To Live Among the Stars: The Odyssey to Build an Artificial Environment* (2020). His website is www.worldoftrons.com.

Susan Elizabeth Swanberg, M.S., M.A., J.D., Ph.D., is a former bench scientist and currently assistant professor of journalism at the University of Arizona in Tucson where she teaches reporting, science journalism, environmental journalism, and media law. Her research interests include journalism's historical role in disseminating scientific information to the public; science journalism, propaganda, and the nuclear age; and vetting scientific expertise in a post-truth society.

Index